ON THE
BURNING
EDGE

DISCARDED

ON THE BURNING EDGE

A FATEFUL FIRE
AND
THE MEN WHO
FOUGHT IT

KYLE DICKMAN

 BALLANTINE BOOKS NEW YORK

Published in the United States by Ballantine Books,
an imprint of Random House, a division of
Penguin Random House LLC, New York.

BALLANTINE and the HOUSE colophon are registered
trademarks of Penguin Random House LLC.

ISBN 978-0-553-39212-8
eBook ISBN 978-0-553-39213-5

Printed in the United States of America on acid-free paper

www.ballantinebooks.com

9 8 7 6 5 4 3 2 1

FIRST EDITION

Book design by Barbara M. Bachman

For Turin

CONTENTS

DRAMATIS PERSONAE

THE GRANITE MOUNTAIN HOTSHOTS

Start of the 2013 season;
in order of experience accrued

OVERHEAD

Eric Marsh, superintendent

Jesse Steed, captain, acting superintendent for
majority of 2013 season

Tom Cooley, acting captain

Travis Carter, squad boss in charge of saw teams

ALPHA SQUAD

Clayton Whitted, squad boss, former youth pastor

Christopher MacKenzie, lead firefighter,
Donut's roommate

Brendan "Donut" McDonough, lookout on
Yarnell Hill Fire

Scott Norris, sawyer, former Payson Hotshot

Anthony Rose, second-year hotshot

Joe Thurston, Norris's swamper

Grant McKee, rookie, youngest hotshot on crew

Renan Packer, rookie, McKee's "battle buddy"

John Percin, Jr., rookie

BRAVO SQUAD

Robert Caldwell, squad boss, McKee's cousin

Travis Turbyfill, lead firefighter

Andrew Ashcraft, lead sawyer

Brandon Bunch, fourth-year sawyer, Janae's husband

Wade Parker, Bunch's swamper

Garret Zuppiger, second-year hotshot

Dustin DeFord, rookie

Sean Misner, rookie

William Warneke, rookie

Kevin Woyjeck, rookie, aspiring structural firefighter

KEY YARNELL HILL FIRE PERSONNEL

Todd Abel, operations chief

Rogers "Trew" Brown, Blue Ridge Hotshots captain

Gary Cordes, structure protection specialist for Yarnell

Steve Emery, body recovery team member

Truman and Lois Ferrell, Yarnell homeowners

Brian Frisby, Blue Ridge Hotshots superintendent

Rance Marquez, Division Zulu

Paul Musser, operations chief

Eric Tarr, medic

Darrell Willis, structure protection specialist for
    Peeples Valley

I f Eric Marsh had any doubts about sending one of his firefighters alone into the path of a raging wildfire near Yarnell, Arizona, the veteran superintendent didn't speak them out loud. He certainly had reason for concern. Marsh was in command of the twenty-man Granite Mountain Hotshots, and on June 30, 2013, the Yarnell Hill Fire was bearing down on the 650-person town. Already the blaze had defied the expectations of the few hundred firefighters from half a dozen different agencies on scene. Ranches were burning, a middle school was soon to catch fire, and entire subdivisions of homes were threatened. Marsh's security lay in the fact that all of his men stood just feet away from the safety of the "black"—the already torched brush that couldn't reignite. All of his men, that is, except for one: Brendan "Donut" McDonough.

Marsh and Granite Mountain were working on a long ridgeline above town, and to keep constant watch on the distance between the main fire and the other nineteen men, Marsh had sent Donut into an unburned valley that, unbeknownst to him, was soon to be consumed by flames. By 3:30 P.M. the fire's three-mile flank was still smoldering a few hundred yards north of Donut's lookout position. Marsh knew the flank wasn't likely to remain so well behaved. The forecast was calling for thunderstorms, along with forty-to-fifty-mile-per-hour gusts out of the north—a threefold increase over the steady breeze that had fanned the flames all day.

Marsh felt the wind before most other firefighters and immediately radioed his supervisors to warn them of the dangerous shift. Minutes later, the long flank jumped to life like a magician's trick. Along a three-mile line of brittle chaparral brush, what had been a veil of wispy white smoke transformed into a uniform wall of thirty-foot flames now advancing toward the homes in Yarnell. Evacuations hadn't started, but when the wind shifted, homeowners began frantically loading valuables into their cars. Not even the dozens of vehicles on scene—fire engines and a DC-10 air tanker, a specially modified commercial jet that could unload thirty thousand gallons of fire retardant in a single drop—could stop the flaming front. The town had only hours to stand.

Marsh watched the flames leap unchecked across a drainage about a quarter-mile from Donut's lookout. Then he heard Donut's radio call to the crew's second-in-command, Jesse Steed.

"Steed, Donut." *Heavy breathing.* "It's hit my trigger point. I'm heading back to the safety zone."

Donut dropped off the back side of the granite knoll and started sprinting through brush so dry it cracked as he swatted it aside. Embers and ash drifted over his head. He plowed through the bushes toward a small clearing around the metal skeleton of a long-abandoned road grader.

As he stumbled into the clearing, Donut suddenly realized it wasn't possible to outrun the blaze—a nightmare scenario for every wildland firefighter, made worse by the fact that the clearing, his only chance for survival, was barely the size of a tennis court. He couldn't outrun the fire. Flames had halved the distance between the drainage and the clearing in minutes. The wind thrashed the walls of scrub oaks and manzanita bushes surrounding the grader. Even if Donut deployed his fire shelter, an aluminum shield designed to deflect heat, his chances of surviving a burnover were grim. Marsh knew as much, too.

The emergency lights of fire engines rushing toward Yarnell lit up the distant horizon. Air tankers dropping retardant on the blaze swarmed overhead, and somewhere on the ridge near Marsh, the nineteen other members of the Granite Mountain Hotshots watched from the safety of the already blackened fuel. None of them could do anything to help Donut. Marsh was about to watch flames overrun his crew member—then an entire town.

Less than an hour later, Yarnell was ablaze, and Donut was the only surviving member of Granite Mountain. Marsh, Steed, and seventeen other hotshots were all dead. Inexplicably, the crew had left the black to move to the safety of a ranch closer to town, and the fire had trapped them in the no-man's-land between the two safety zones. The deaths of the nineteen Granite Mountain Hotshots represented the most firefighters killed in the line of duty since 9/11. It was the single greatest loss of life among wildland firefighters since the 1930s. For many first responders, June 30, 2013, had the same instantaneous effect as 9/11. They stopped whatever they were doing to try to comprehend the gravitas of the situation. Most wildland firefighters in the nation—current or former—can tell you exactly where they were when they heard the news.

I was on my front porch in Santa Fe, New Mexico. I spent five summers fighting fires in California, and Granite Mountain's deaths came a week after I'd finished a story for *Outside* magazine about embedding with my former hotshot crew during the 2012 fire season. On July 9, *Outside* sent me to Prescott, Arizona, to attend the hotshots' memorial. This was the beginning of a yearlong journey to find out exactly what had happened at Yarnell Hill. The reporting would take me from the mountains of New Mexico to the basin where the men had died, but most often, I returned to Granite Mountain's home base, in Prescott.

I first met Donut more than a month after the tragedy, in a dark office at Prescott's Yavapai College, the local community college. He was twenty-two and rangy, with blond hair cropped close, and wore a tank top and cutoff jean shorts. His leg was red and swollen. That week he'd gotten a new specialty tattoo, above an older piece that blended a frosted doughnut with Granite Mountain's logo.

"It's the most painful tattoo you can get," Donut said, implying that the physical pain had provided relief from something worse.

His phone vibrated every few minutes. Sometimes it was a text from a concerned friend or loved one; more often it was a call from a news outlet—*Good Morning America,* CBS, *GQ,* the local paper— requesting an interview. Donut gave dozens, but his explanation of what happened on Yarnell Hill that day never really fit into sound bites. Donut wasn't the most polished guy to begin with, and he'd only just begun to process the disaster that had killed so many of his friends. Most stories that emerged were less about what had happened during the fire and more a reflection of Donut's grief and incomprehension.

We spoke for two and a half hours that first day. Over the next year, I talked to Donut often as he battled the worst of his posttraumatic stress and came to terms with survivor's guilt. He took me to the locations of fires Granite Mountain fought during the 2013 season, told me stories of the crew, and related what it's like to be a hotshot in an age when fires are burning more intensely than at any time in modern history.

But ultimately, this wasn't a story that could be told through one man's experience alone. Yarnell Hill's greater significance was to be found in the overall context of current and historic wildfires and the hugely destructive 2013 fire season. Dozens of fire researchers and historians, longtime hotshot superintendents, and forest scientists weighed in on the implications of this blaze—and other enormous fires like it—for the future of the American West. These experts also explained why, under current fire policy, more tragedies like Yarnell Hill can be expected.

Donut told me of two other hotshots who had left Granite Mountain just weeks before Yarnell. These three survivors and other lifelong wildland firefighters made it clear that Granite Mountain was different from the 113 other hotshot crews working in the United

States. Much of Granite Mountain's unique personality and culture stemmed from the crew's leader, Eric Marsh. When I first met his widow, Amanda, at their horse ranch outside Prescott, she told me, "People are going to build him up—make him a hero—and then tear him down." Her husband's personal history was colored, but, more significantly, his fire-line record was varied. One hotshot superintendent who'd worked with Marsh called him a "bad-decisions, good-outcome guy," while others considered him among the most skilled firefighters they knew. Marsh emerged as a highly intelligent and deeply complicated figure. He was key to understanding this tragedy.

Through long interviews with surviving members of Granite Mountain, other firefighters who worked closely with the crew, and dozens of the hotshots' family members, as well as careful study of thousands of pages of archived documents left behind after two official investigations into Yarnell Hill and four other significant fires the crew fought in 2013, it became possible to explore the critical questions surrounding the tragedy that will forever haunt wildland firefighting. Who were these men? What were the historical forces that led to the unprecedented blaze? And, crucially, why did nineteen men leave the relative safety of the blackened earth, only to die shortly after in the most horrific manner imaginable?

Fire Season, 2013
Granite Mountain Hotshots
Prescott, Arizona

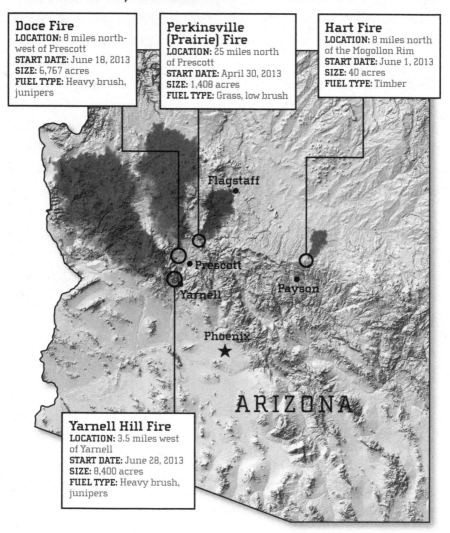

Doce Fire
LOCATION: 8 miles north-west of Prescott
START DATE: June 18, 2013
SIZE: 6,767 acres
FUEL TYPE: Heavy brush, junipers

Perkinsville (Prairie) Fire
LOCATION: 25 miles north of Prescott
START DATE: April 30, 2013
SIZE: 1,408 acres
FUEL TYPE: Grass, low brush

Hart Fire
LOCATION: 8 miles north of the Mogollon Rim
START DATE: June 1, 2013
SIZE: 40 acres
FUEL TYPE: Timber

Yarnell Hill Fire
LOCATION: 3.5 miles west of Yarnell
START DATE: June 28, 2013
SIZE: 8,400 acres
FUEL TYPE: Heavy brush, junipers

Flagstaff

Prescott

Yarnell

Payson

Phoenix

ARIZONA

Thompson Ridge Fire

LOCATION: 10 miles north of Jemez Springs
START DATE: May 31, 2013
SIZE: 23,965 acres
FUEL TYPE: Timber

Jemez Springs

★ Santa Fe

Albuquerque

NEW MEXICO

Maps by John McCauley

PART ONE

TINDERBOX

CHAPTER 1

THE PERFECT PROFESSION

Grant McKee parked his aging Dodge Neon by a row of pickups out front of an aluminum-sided shop in Prescott, Arizona. The air shook with the punch of nail guns and the rattle of diesel engines, and a dry breeze wicked the last of winter's humidity from the desert Southwest. It was early April, and the new leaves on the only two trees inside the fenced-in compound rustled quietly.

From the driver's seat, Grant watched seven or eight men huddle together inside the shop's raised garage door. Like him, nearly all were in their early twenties and fit—prime draftees, had Uncle Sam come calling. They wore oil-stained Carhartt pants, faded flannel shirts, and down jackets with duct-tape patches covering holes. Most had quit jobs washing dishes or parking cars as golf course valets to be frontline grunts in America's long war on wildfires. None of them could have known it that April morning, but the Granite Mountain Hotshots would come to define their lives.

Grant made the long walk toward the men. His unused leather workboots, heavy and foreign, clicked on the blacktop as he entered the garage. He had an angular face, pronounced eyebrows, and dark hair recently cut to give him the put-together look he liked. But what caught the other firefighters' attention was his build: five feet seven inches, with the same muscled frame that had carried him to the varsity wrestling team as a high school freshman.

Inside, most of the guys silently sized the others up. Some knew one another. Others looked terrified. One fidgeted nervously, as if just being in Granite Mountain's hallowed garage fulfilled a dream. Grant didn't feel that way. He might have looked like a hotshot, but 2013 was his first season, and he wasn't sure he wanted to be one. If all went as planned, he'd make $30,000 by November and, with a résumé padded with first responder experience, move on to the paramedic job he actually wanted. At first glance, wildland firefighting seemed an odd step for Grant, even if the City of Prescott's fire department hosted the wildland fire crew. He preferred movies and comfortable beds to the outdoors, and over the next eight months, the job would afford him very little time inside.

Granite Mountain's modest and nondescript headquarters took up a corner of the closest thing the forty-thousand-person city of Prescott had to an industrial district. It didn't look much like a real fire station. No hoses, no turnouts, red trucks, or spotted dogs. The inside smelled of two-stroke gasoline and stale woodsmoke, and the walls were decorated with some of the more interesting things the guys had found in the woods. A deer's jawbone sat on the tool rack with the screwdrivers, and a dozen rusted-out pull-tab Coors cans with sun-cured pine needles sticking out of the tops filled one shelf. A jumble of wooden tool handles, some splintered from crushing swings, filled an entire corner, and chainsaws and their parts—chains, carburetors, air filters—cluttered a rolling island in the middle of the room. The hotshots called the place the saw shop, and Grant and a dozen other firefighters milled about, waiting for the shift to start.

A few minutes later, a diesel Ford F-250 that was lifted eighteen inches off its oversize tires rumbled into the parking lot. Donut emerged and walked straight up to Grant and the other rookies. "You guys ready to learn?" he asked. Then he took his place nearby and struck up a loud conversation with other veterans.

"Get your asses in here!" boomed Eric Marsh, the crew's superintendent, from inside the station's ready room. He always looked forward to the return of the seasonal firefighters. Having fourteen more guys around—there were no women on this year's crew—ensured a livelier atmosphere around the station.

Grant joined the throng of men funneling through the shop door and took a right into a conference room. A dozen Formica tables and folding chairs sat before blank whiteboards. Hanging on the walls were fire safety reminders—YOUR LIFE IS MORE IMPORTANT THAN ANY STRUCTURE!—and posters memorializing the two deadliest wildfires in recent history: Montana's 1949 Mann Gulch Fire (thirteen deaths) and Colorado's 1994 South Canyon Fire (fourteen deaths). Up front, Marsh sat alone.

"Morning, gentlemen," Marsh said. A seasoned firefighter, he was in his early forties, a handsome man with a lantern jaw, a handlebar mustache, and a red-and-black Granite Mountain Hotshots ball cap pulled over a head of salt-and-pepper hair. One arm was in a sling from a mountain-biking accident weeks earlier, a fact he chose not to elaborate on.

As the shuffling of chairs quieted, Marsh looked out at the faces. Five crewmen had left Granite Mountain over the winter, and Marsh had to hire replacements, which was always a bit of a gamble. Hotshot work is rarely glamorous, and the months away from home, brutal physical labor, heat, poison oak, chainsaws, falling trees, and flames aren't for everybody. Usually, a few new hires washed out in the first month or two, a few made it through a season but swore to never return, and a few fell in love with the job.

It was good to see some of the veterans back. Some of the hotshots had worked as seasonal employees under Marsh since he took over as Granite Mountain's superintendent in 2006. There was Brandon Bunch, the laconic former bull rider who was starting his fourth season with Granite Mountain, and Andrew Ashcraft, an active member of the Mormon church, who was still sporting his well-groomed mustache. At twenty-nine, Ashcraft was tall with the square-jawed good looks, dark hair, and swagger of Tom Cruise in *Top Gun*. He also had four kids. Marsh was in the process of converting him to a permanent employee, but for the time being, the paperwork was caught in bureaucratic limbo. For the third year, Ashcraft was officially returning as a seasonal employee.

Marsh introduced himself and asked the crew's full-time firefighters to do the same. Known collectively as the overhead, these seven hotshots commanded Granite Mountain. Jesse Steed, thirty-six, was the captain, the number two in command. People called Jesse the picture-perfect hotshot, and the ex-Marine, who was six-four and more than 220 pounds, never disputed the point. Beside Jesse sat the crew's three squad bosses: Travis Carter, thirty-one, who was a walk-on football player at the University of Arizona and still looked the part; Clayton Whitted, twenty-seven, a former youth pastor who brought church to the fire line; and Robert Caldwell, twenty-three, a Prescott local with an IQ high enough to merit his acceptance in Mensa. The final two overhead were lead firefighters Chris MacKenzie, thirty, a California-born longtime hotshot with such a laid-back demeanor that his crewmates often compared him to the Dude in *The Big Lebowski,* and Travis Turbyfill, twenty-seven, a moose-size man who had seen combat in Afghanistan as a Marine and was a gifted mechanic.

"Tell us about yourselves," Marsh said to the crew. Most of them didn't know one another and didn't want to speak. Marsh's way of dealing with their shyness was to force his men to get over it. "Give us your name, how many years you've fought fire, and something . . . your favorite color, whatever you'd like," he said.

At twenty, Grant was the youngest guy on the crew. He played it cocky, standing, saying his name, and coolly explaining that his cousin was Bob Caldwell, one of the squad bosses. It didn't go so easily for all the new guys.

"Dustin DeFord," said a redheaded rookie who sat near Grant. "My favorite color is"—DeFord paused for a second—"black. It reminds me of fire."

"That's nice, Dustin," Marsh said. "Thank you. Now please sit the fuck down."

———

There are 114 twenty-person hotshot crews nationwide, but the vast majority are stationed in the West. In New Mexico and Arizona, there are a total of twenty. With a forty-one-year-old U.S. Forest Service crew stationed on the town's north end, Prescott has two. Hotshots make up less than 5 percent of the estimated fifty-six thousand federal and state wildland firefighters who battle blazes every summer. The remainder work on engines, less-qualified crews, water tenders, and aircraft, and as support staff. While the media often likens hotshots to the special forces of wildland firefighting, it's an imperfect comparison. Unlike joining the Navy SEALs or the Army Rangers, which demand years of training, hotshot crews can accept new hires within a matter of days. On paper, becoming a hotshot requires only two basic courses, which can be completed online in hours, and the physical ability to hike three miles in forty-five minutes while carrying a forty-five-pound pack. What separates hotshots from the other wildland firefighters is that they specialize in fighting America's largest and most dangerous blazes.

No job in the firefighting profession requires its men and women to spend so much time on the edge of an active burn. Over the course of a thirty-year career, a hotshot might fight eight hundred to one thousand fires. By contrast, a structural firefighter may battle only eighty over that same period. At its best, the job is the greatest adventure of many hotshots' lives. At its worst, it varies little from the hard labor of a chain gang. For weeks, crews work sixteen-hour days in soaring temperatures, using chainsaws and hand tools like Pulaskis and shovels to build control lines, or linear barriers clear of flammable materials, around wildfires. There are female hotshots, but usually no more than one or two on a given crew, and the overwhelming majority are young men in their twenties and thirties. Some are college-educated, having quit high-paying finance jobs in search of a simpler lifestyle in the woods, while others have left behind farms, reservations, or the inner city to make a year's salary in eight months of work.

Crew cultures tend to be militaristic. Among the standard twenty-head crew, at least seven are highly trained permanent firefighters—the overhead. Among them are emergency medical technicians, men and women with decades of fire-line experience, and chainsaw operators capable of cutting down trees as thick as some bridge pilings.

During the first few weeks of fire season, the overhead lead the new hotshots through a series of runs, hikes, and workouts that get the men into the best shape of their lives. Fitness is a matter of safety on the fire line: It's required for moving efficiently through the mountains, but it's also mandatory should a firefighter have to outrun a blaze. Granite Mountain called this intensive training period the two-week critical, and it culminated in a drill at the end of April meant to test the hotshots' new skills and fitness.

A week into the 2013 season, Marsh gathered the men and told them about his shoulder. He couldn't return to the fire line until it healed, which might take months. The news meant that for the first time in the decade that Marsh had been a part of the crew's overhead, he wouldn't take part in the drill. His injury had been too severe. He made no effort to hide his disappointment. Marsh loved being superintendent and fighting fires. Until he returned, Tom Cooley, a City of Prescott structural firefighter with many years of hotshot experience, would take over as captain, and Jesse Steed, officially the captain, would become the interim superintendent.

The news thrilled Steed. He talked openly and often about his plan to someday take over as Granite Mountain's superintendent, and even the temporary position gave him a chance to shape the crew. Among the first things Steed did was make the physical training punishingly intense.

The veterans expected him to do as much. During slow shifts on the line, Steed would often pound out a few dozen squats with a chainsaw on each shoulder or do lat pulls with forty-gallon buckets of foam—the soapy substance, mixed with water, that's used to slow a fire's spread. He designed the hardest workouts himself. One day he led the men to a three-hundred-foot hill and made them run up in their fire boots. Then they'd sprint back down, grab a chainsaw, and repeat the "Circle" six times. The hike up Thumb Butte, an aptly named chunk of black basalt that rises above Prescott, was even harder. Steed paired the hotshots up and made the teams race the two miles and six hundred vertical feet to the top, one man carrying the other on his back.

Most days, the intensity of Steed's workouts made two or three guys puke from overexertion. Grant pushed himself so hard he vomited every day. In many ways, the work didn't suit Grant. He was stylish, clean, and fastidious. Every night before bed, he'd fold and stack his firefighting clothes and place them at the foot of his bed in the order he'd put them on the next morning. Into his shoes went his socks—one designated for the right foot, one for the left—and those went next to the toilet for quick entry in the morning.

It had been Grant's cousin, Bob, who suggested he apply to be a hotshot. "Think of it as a stepping-stone," Bob had told him. "Put in a few years on a badass crew like Granite Mountain and down the road it'll help you get a job on an ambulance or engine."

Charming and easy with people, Grant interviewed well. Even so, Marsh had hired him only after one of his top candidates turned down the job. Grant easily completed the three-mile pack test but barely met the crew's fitness requirements, which were many steps harder than the basic national standard. Applicants had to run a mile and a half in less than ten minutes and do seven pull-ups, forty sit-ups, and twenty-five push-ups—each series in less than a minute. Once a fiercely fit wrestler, Grant was still in good shape and had recently run a half-marathon, even stopping to smoke cigarettes on the way. With that foundation, Grant believed he could simply will his way through whatever challenges hotshotting threw at him.

In the first week of training, Steed disproved Grant's naïve theory. In between classes on how to correctly run a chainsaw, sharpen a tool, or take weather measurements on the fire line, the crew spent hours working out on the sweltering blacktop inside the chain-link fence surrounding the station's compound. On any given day, they might run six miles, hike three, and do 60 pull-ups, 390 sit-ups, and 240 push-ups. Steed, who always led from the front, seemed to enjoy the men's suffering.

Sometimes, between their tenth and twelfth set of twenty-five push-ups, he would call out "Hold!" and the men would pause with their bodies prone and their noses just an inch above the ground. Grant quaked, his whole body trembling in ugly, uncontrollable shudders. Just a week into the season, he already felt poisoned by lactic acid. Bob promised him that he'd get used to the workout and the training would get easier. It'd have to, Grant thought, because if it didn't, he'd quit.

CHAPTER 2

TRAINING DAY

Just a few weeks into the season, Granite Mountain's intensive training concluded with a live drill meant to test the hotshots' ability to adapt to the constantly changing environment of a firefight. To ensure competence, all fire engines, hotshot crews, and air tankers—known, as all fire assets are, as "resources"—must complete such a drill every year. The City of Prescott's Wildland Division chief, a gap-toothed and devout man named Darrell Willis, would judge the hotshots. If they passed, Granite Mountain would be certified to fight fires for the 2013 season. At 7 A.M., an hour earlier than usual, the men came to work and left the station immediately after. As they usually did, the crew's vehicles headed out in a certain order.

Steed led the way in the superintendent's truck, a Dodge Ram 4500. A hotshot rode with him in the passenger seat. In the years before, Marsh had customized Granite Mountain's "supe truck," as it was known, by welding taillights that cast the letters GMIHC, for "Granite Mountain Interagency Hotshot Crew." Following closely behind Steed were a second small truck, called the saw truck, that carried Travis Carter, the squad boss, and another hotshot, and finally the two buggies, burly ten-person crew carriers with oversize tires and enough storage space to carry the equipment needed to support twenty men for two weeks at a time. The caravan of white-and-red wildland fire vehicles was a familiar site in downtown, and most locals didn't give the trucks a second glance as they rumbled past the courthouse, and the 127 stately elms surrounding it, at the city's center.

Prescott (pronounced by locals to rhyme with "biscuit") sits in the pine forests an hour and a half north of Phoenix, but after decades of rapid population growth, it's no longer the strictly blue-collar town it was for much of the twentieth century. The town now has a small aerospace industry and three colleges within city limits. Hip new microbreweries and boutique coffee shops are pushing out some of the ubiquitous antiques stores, but Prescott's defining characteristic remains its western pride. Kids here grow up dreaming of becoming bull riders and spend their summers working on cattle ranches that operate in the rangeland outside town. Many locals have roots that reach back to the city's 1860s founding, when miners struck gold in the surrounding Bradshaw Mountains, and for a short period the town served as the territorial capital of Arizona. The Earp family and Doc Holliday stalked Prescott's streets. It's on Whiskey Row, a half-mile strip adjacent to the courthouse that once boasted forty bars, that the Old West feels most present.

That morning, as the hotshots headed out of town for their drill, the elm-shaded sidewalks of Whiskey Row were empty except for a few early-season tourists taking in displays of cowboy boots and faux-Indian headdresses. Flyers touting the 126th anniversary of the world's oldest rodeo were plastered onto the windows of the hole-in-the-wall bars that are still packed into the strip, and the buggies' diesel engines echoed against the brick facades.

Grant peered out of the buggy's submarine windows as Granite Mountain's caravan passed from Whiskey Row south into the wilder lands of the Prescott National Forest. Grant watched the housing styles change from historic Victorians to seventies-era ranch homes to double-wide trailers and finally to McMansions perched in the foothills. Just a few miles from downtown, the buildings vanished altogether. The roads, once paved and suburban, became dirt or gravel, and ponderosa pines filled the view. On the oldest trees, the bark was blackened from past fires. The hotshots, drowsy and still unaccustomed to the day's early hours, watched the pines flicker by. Most of the men, nervous about the coming test, mentally rehearsed the skills they'd learned over the previous weeks and focused on their assigned roles. In the stillness, Grant saw an opportunity to ease the tension.

"Hey, Chris!" he called to the cab from the dim light in the back. "What are we going to do today?"

"You gonna learn, boy!" Chris, the lead firefighter, yelled back to his squad from the buggy's cab. He'd said the same thing before every workout since day one. At first, Chris's yelling terrified the rookies; now the joke was theirs, too. Anytime Chris forgot to deliver his daily message, Grant reminded the lead firefighter.

From basic crew structure to the nation's coordinated response to wildfires, nearly everything in wildland firefighting is organized hierarchically. There's generally one person or organization in charge, with layers of command cascading down from the apex. It was no different on Granite Mountain.

After Marsh was injured, Steed claimed the crew's top position. Tom Cooley, the temporary captain, was just beneath him. The rest of the hotshots were divided into two modules, called Alpha and Bravo. Each squad of either nine or ten hotshots was assigned a buggy and contained a mix of veterans and rookies. Command of the smaller units fell to their respective squad bosses: Clayton Whitted on Alpha and Bob Caldwell on Bravo. (Travis Carter, the third squad boss, ran both Alpha's and Bravo's chainsaw teams.) Below Clayton and Bob were two lead firefighters—Chris on Alpha, Travis on Bravo—who could step into the squad boss role; a pair of hotshots known as sawyers, who ran chainsaw; and another pair who removed the vegetation the sawyers cut: the swampers. Rounding out the squads were the ten or eleven firefighters whose job was to use hand tools to dig fire line. These firefighters are called the scrape. At any point, a hotshot might be instructed to run saw, dig line, or swamp, and one of the nine qualified veterans might serve as a lookout posted to warn the crew of any unexpected change in fire behavior, but otherwise the men defaulted to their assigned roles.

Among seasonal hotshots, the scrape is considered the bottom of the hierarchy. It's where nearly all rookies start their fire careers. Over the course of the season, the scrape try to prove themselves capable of becoming swampers, while the swampers are trying to move up to the sawyer job, which is the most sought-after position among seasonals. Grant was in Alpha's scrape, and since he was a rookie, the seat the veterans assigned to him was in the rear left, the bumpiest of the eight seats in the back.

He was the first out of Alpha's buggy when the caravan stopped at a dirt parking lot walled in by 150-foot pines. The morning was still cool, with traces of night's humidity still in the air. The hotshots, wearing yellow fire-resistant shirts, green pants, and black hard hats, gathered around Steed. He and Marsh, who had come to supervise the drill, wore the same uniform as the men, but their helmets were red.

Marsh stood quietly to the side and took notes on the drill while Steed spread out a map on the hood of his truck and pointed out the fake fire's location. It sat in a saddle of the Bradshaws between the Senator Highway, a now-decrepit stagecoach road from the 1860s, and Highway 89, which runs south through the town of Yarnell and toward Phoenix. The flag fire, marked before the drill started with strips of plastic pink flagging, was a few acres and spreading quickly. Steed identified for the men their escape routes—clearly demarcated paths of retreat—and safety zones: areas cleared of flammable materials that are large enough for firefighters to weather an out-of-control blaze. Sometimes safety zones are clearings bulldozed into the "green," the unburned vegetation beside a blaze. But more often, when a fire explodes, hotshots retreat into the cold ash left behind by the flames—the "good black." With no chance of rekindling, it's often the safest place to be during a rapidly intensifying fire. On the flag fire, Steed pointed inside the theoretical blaze's perimeter. Granite Mountain's safety zone was the good black.

Wildland firefighters bring blazes under control by building boxes of nonflammable things around them. In one combination or another, that means surrounding the flames with water, roads, rock, bare dirt, and already burned fuel. To do this, firefighters have a relatively short list of tools at their disposal. Engines, air tankers, and helicopters use water and retardant to either knock down the fire's spreading head or soak the vegetation ahead of the flames to slow its growth. Hotshot crews, though, use chainsaws and hand tools to remove the vegetation and create a continuous line of bare dirt or rocks around a fire.

The guiding principle of fire agencies is that the cheapest and most effective way to fight fires is to catch them when they're small. They're kept small by attacking them early. In a typical blaze, one of the country's 826 operating fire lookouts—or, these days, someone with keen eyes and a cell phone—calls in a smoke report, and the closest available resources mount the initial response. The first wave of firefighters sent to a new blaze during peak fire season is usually designed to be overkill. On a hot and breezy June day, that first order might include three fire engines, a twenty-person handcrew, a helicopter, and an Air Attack plane that does what ground resources cannot by tracking the fire's overall progression from above.

Engines, limited by road access and the amount of water they can carry, can hose down the flames faster than hotshot crews can build line. As such, in hopes of slowing the fire's spread, engines tend to lead the initial attack while the hotshot and handcrews follow behind. If the flame lengths are small, crews build lines directly on the fire's edge. This is called "going direct," and if crews succeed in lassoing the blaze, the tactic stops the fire where it lies. When a fire's burning more intensely, hotshots step back anywhere from a few hundred yards to five miles and build "indirect line" by constructing firebreaks on ridges ahead of the combustion. Once the box (which is rarely actually square) is complete, hotshots burn out or back-fire, intentionally igniting the vegetation between the line and the wild flames, thereby robbing the blaze of the fuel it needs to survive.

During the drill in the pine forest outside Prescott, the men of Granite Mountain needed to prove that they were capable of using all the skills needed to control fires of varying sizes. With Steed's orders in hand, they broke into action.

Heavy aluminum cabinets creaked open. On one side of the buggy men found their Pulaskis—ax-adze combinations that are the signature tool of wildland firefighters—and rhinos, a burlier version of a hoe. On the opposite side, the sawyers unloaded their chainsaws while the swampers passed out red liter bottles of extra gas and oil to each crewman. On a busy day, a sawyer can go through more than twenty liters of fuel, and all the hotshots shared the burden of the extra weight. Within minutes, Granite Mountain was ready to go. With food, a jacket, files for sharpening tools, parachute cord for tying up whatever needs to be tied up in the woods, flares for lighting backfires, a lighter, a file, spare chains or chainsaw air filters, eight quarts of water, and half a dozen other random items, each man's pack weighed forty-five pounds or more. Then there were the things they carried in their hands.

As the hotshots scurried around the buggy, one of the senior firefighters plunked down a five-gallon plastic-lined cardboard box of water at Grant's feet. The cubie, as it was called, weighed another forty-five pounds. Steed and the squad bosses knew the hotshots wouldn't need the additional water that day. Grant was given the cubie to prove that he could handle the extra weight.

Grant threaded his tool through the cubie's plastic handle, hauled the box to his shoulder like a bindle, and took his place in the single-file line that had formed behind Steed. The sawyers hiked closest to Steed, with the swampers following behind and the scrape toward the back. A squad boss took up the rear and scolded any hotshot who slowed down, to keep the long line of men from pulling apart and squeezing back together.

Steed hiked faster than most people jog. His strides were long and steady, his weight slightly forward and centered over his knees. For twenty minutes, the crew hammered uphill along a two-track dirt road. Soon, sweat soaked through the men's yellow shirts and the hotshots heard nothing more than the sound of their own breathing.

Not long into the hike, Steed's oppressively fast pace broke the rookie John Percin, Jr. He'd wrenched his knee during training and pulled up lame during the hike. One of the squad bosses bawled at him to catch up to the others, but Percin balked. The hike proved too painful. He dropped out, and a lead firefighter led him back down the hill to the trucks. Percin's knee would take more than a month to heal, and he wouldn't work with the hotshots again until the last week of June.

The rest of the crew didn't pause. They kept racing to their objective: the series of pink plastic strips of flagging tied to tree branches to mark the fire's perimeter. Grant, with the extra weight of the cubie, was struggling to keep up, and finally he, too, peeled off from the single-file line and dropped to his knee, gasping for air. With his face inches above the sun-baked pine needles, their sweet vanilla smell drowned out by the sheer volume of air sucking into his nostrils, Grant returned to the question he'd been asking himself for weeks now: *What was I thinking volunteering to chase a fitness-obsessed ex-Marine up the side of a mountain?*

Grant's goal was to help people as a medic—not save forests that he wasn't particularly fond of anyway. He'd camped only three times, ever, and found the experience to be little more than an exercise in discomfort.

Grant grew up a latchkey kid in Orange County, California. His home life was rocky and he started taking pills in high school. Unhappy with his life's direction, he moved to Prescott as a sophomore in high school to live with his aunt and uncle, Linda and David Caldwell, and his older cousins, Bob and Taylor. Still, even in a more stable home, it took Grant a couple of years to straighten out. He had to repeat his junior year of high school. Eventually, he stopped abusing pills and even went on to use his own experience to teach Drug Abuse Resistance Education. After graduating, he started taking EMT classes at Prescott's Yavapai College and fell in love with Leah Fine, a pretty blonde he'd met through a friend and who had a shared interest in running.

Leah fell for Grant's sincerity. His unabashed love for her was like nothing she'd ever experienced, and his honesty about emotion was raw and endearing. They moved in together and talked often of eloping in the fall of 2015. Leah was far more outdoorsy. She liked sleeping under the stars and craved exercise. When Grant first signed up to be a hotshot, they joked that she was far and away the better candidate for the job. Instead, Leah worked as an assistant at a property management firm and put her aspirations of going to college to be a photojournalist on hold. Before going to work, she made Grant salads for lunch, which otherwise would have been gas station burritos and maple bars, and when he came home she would coddle him as he complained about the punishing workouts that Steed dished out every day. In the beginning, he'd come home to her threatening to quit. But Leah never doubted Grant's toughness. She was less certain that he could handle the time away from her.

"For the first time in his life he actually had a place he felt he belonged," Leah said. They shared a comfortable rental with their dog, Derek, and a few roommates. His job bartending and washing dishes at a local Mexican restaurant was the only hitch in what Grant otherwise considered the smoothest period of his life.

When Bob first suggested that Grant apply for Granite Mountain in the winter of 2012, the prospect wasn't appealing. Leaving Leah for any amount of time, let alone many months, sounded awful to him. So did sleeping in the dirt. Ultimately, though, the lure of professional job experience and a chance to earn a year's salary with just eight months of work trumped Grant's love of creature comforts. Plus, Leah pushed him to take the job. Her encouragement made it easier to believe that hotshotting, no matter what challenges the job presented, couldn't possibly be worse than manning a sink.

"Get your ass up! Don't you fucking quit!"

Back on the side of the mountain, Scott Norris, a wiry twenty-eight-year-old redhead who'd transferred to Granite Mountain from a Forest Service hotshot crew that spring, pounced on Grant before he could catch his breath. It was usually the squad bosses' job to discipline the rookies, but Scott, who'd wanted to make a career of firefighting, took it upon himself. A few of the guys marching behind Steed glanced back. Grant was still down on one knee, but now he was vomiting again.

Kneeling on the ground, his vision blurred by fatigue, with Scott yelling at him and the other hotshots still racing up the hill, Grant managed to stumble to his feet. Becoming a hotshot might not have been his dream, but backing down from a challenge wasn't in his nature. Grant gritted his teeth, hauled the box of water to his shoulder, and chased after the rest of the crew.

Marsh and Steed had designed Granite Mountain's training day with a larger purpose in mind. The crew would build a line that, later in the fall or winter, would serve as a line meant to contain a prescribed burn. The eventual fire would accomplish two objectives. First, it would mimic the natural low-intensity blazes that were common to the Prescott area before fire suppression took hold as the dominant land management policy, in around 1910. And second, the burn would check any future wildfires by leaving something like a moat of blackened forest around Prescott.

The prescribed-burn plot needed to be lined at some point that summer, and Steed intended for Granite Mountain to tackle the project with the same urgency they would if the flames were already marching down Whiskey Row. He ordered the saw teams—the sawyers and swampers—to clear a thirty-foot-wide swath of trees, brush, and limbs on the edge of the pink flagging. The scrape would follow behind, digging a ten-foot-wide path down to mineral soil in the middle of the swath. When complete, it would look as if a bulldozer had plowed through the forest and a trail crew had built a jeep road through the middle of the swath.

Scott started working. During the five seasons he'd spent fighting fire with the nearby Payson Hotshots, a crew based a hundred miles to the east, he'd learned to run a chainsaw well. When he came to Granite Mountain, he expected that Steed would bump him back to the scrape with the rookies. Regardless of prior experience, on the Payson crew, new guys always swung hand tools.

Running chainsaw is the most dangerous job on the line. Hundreds of men have knocked out their front teeth while traversing mountains with saws on their shoulders; many have burn scars on their necks from where the muffler, hot from running all day, touched an exposed piece of skin. Sawyers wear Kevlar chaps, eyewear, and gloves to protect themselves in the event that they slip, but on every crew in the nation, somebody has a story of a chainsaw injury. Some are gruesome. Some are life-threatening. It doesn't take much imagination to conjure what happens when some unfortunate hotshot touches a spinning chainsaw.

Many crews won't allow new hotshots, even those with fire experience, to be sawyers until they've proved themselves fit, capable, and safe. It wasn't long after watching the fluid and confident way Scott ran chainsaw that Steed offered him one of the crew's lead sawyer positions.

Scott tugged his machine to life and set it to idle beside him while he pushed in earplugs and pulled his hands into leather work gloves he'd turned inside out to prevent blisters. Joe Thurston, a thirty-year-old Utah native and father of two, had been assigned to be Scott's swamper, and he followed Scott in to work. There was elegance in the way the team moved.

Unable to talk over the screaming engine, Joe occasionally touched Scott's shoulder to alert his sawyer of an overlooked limb. Otherwise, the men, ever aware of the other's presence, seemed to dance around each other. As he grabbed cut limbs, Joe always left a couple of feet between his hands and Scott's saw. At a manzanita bush a little smaller than a Volkswagen Beetle, Joe pulled the branches back to make it easier for Scott to see. The sawyer lopped the limbs off inches above the ground. When one dropped, Joe and Scott made eye contact, and Scott held the saw still for an almost imperceptible moment. In that pause, Joe reached toward the still-whirling chain, grabbed the branch, and tossed it off the fire line. It took seconds to clear the manzanita. Falling trees took longer.

At a larger pine that sat in the middle of the fire line's path, Scott first removed the limbs, running the saw up and down the eighteen-inch-diameter trunk in one fluid motion. Then he hugged the trunk and looked up, assessing the tree's natural lean and determining which direction he wanted it to fall to make Joe's job of removing it from the line easier. Once decided, Scott aimed the pine where he wanted it to go by cutting the trunk perpendicular to its lean. Eight inches above that, at a forty-five-degree angle, he made a second cut that intersected the first. With the toe of his boot, Scott kicked out of the pine a chunk of wood roughly the size and shape of a large watermelon wedge and, still on his knees, spun around to the opposite side of the tree.

"Back-cutting!" Scott bellowed. Travis Carter, the squad boss in charge of the sawyers and swampers, cast a quick glance in Scott's direction to ensure the other three saw teams were a safe distance away from the tree, and Scott checked to make sure Joe was safely behind him. Then, satisfied, he revved the saw and made a flat cut on the back side of the tree. Wood chips bounced off his wire mesh goggles and clung to beads of sweat and the stubble of his red beard. On trees that he was trying to fell against their natural lean, Scott would have Joe hammer a shoe-size plastic wedge into the back of the tree with the small yellow-handled sledgehammer the swamper sometimes carried in the back of his pack.

But wedges weren't necessary on this tree, and the deeper Scott cut, the more the pine leaned over a strap of wood—a hinge—he'd intentionally left in the pine's interior. Then it cracked, and Scott stepped back as the tree broke off the stump and crashed to the forest floor. Every hotshot on the line could feel the ground shake. Scott didn't pause. He strode over to the fallen tree and bucked it into smaller pieces that Joe was already tugging off the line.

Wildfires occur in every state in America. Minnesota's Boundary Waters, Maine's North Woods, Iowa's cornfields, even what little patches of wildland still exist on Long Island: At one point or another, all of it has probably burned, and most of it will burn again. Every few years, Alaska's endless wilderness torches with such intensity that the fires uproot century-old trees. These behemoth blazes can generate vortices of swirling wind, flames, and smoke so powerful that they carry half-ton logs thousands of feet into the air. Somehow fire finds the conditions it needs to thrive in the swamps of Florida's Everglades and the bogs of Louisiana. The natural process of flame and regrowth is many millennia old. For four hundred million years, fuel, oxygen, and heat have combined to make flame.

Americans have been aggressively putting out wildfires only since the beginning of the twentieth century, but in that time we've invented firefighting tools as varied as the landscapes where they're used. Recently, some of the tools have become higher-tech: There are oversize four-wheel-drive fire engines with water turrets that drain twenty-five-hundred-gallon tanks in minutes, helicopter-borne torches that can set ablaze thousands of acres of grasslands in a single pass, remotely piloted drones used to scout potential fire lines without endangering life, and infrared cameras mounted on small planes to map a fire's perimeter and intensity.

But a hotshot's work depends on the relatively simple tools he carries in his hands, and there's one tool every hotshot and wildland firefighter in the country knows intimately: the Pulaski. In 1910, the thirty-five-hundred-person timber town of Wallace, Idaho, sat in the path of a blaze that would grow to the size of Connecticut. Ed Pulaski, a Forest Service ranger, had been tasked with keeping the flames out of Wallace. At one point, he commanded fifty farmers, fathers, and drunks enlisted from the town's doorsteps and bars. Few, if any, knew how to fight fires. When the wind rose and the fire exploded, flames trapped Pulaski and his crew in a canyon. Against their screams and protests, the ranger forced his men into a mineshaft and blockaded the entrance with his body. To run was to die. Pulaski drew his .44 revolver and said, "The next man who tries to leave the tunnel I will shoot."

The militia of firefighters lay down beside one another like sardines, sucking in the cool air puddled on the bottom of the mineshaft. Orange light cast by the fire sprung into the darkness, and it grew so hot, the wooden joists and braces caught flame. Pulaski lost consciousness. When the fire passed and the first of the forty-five or so survivors crawled over their leader's lifeless body, someone said, "Come on outside, boys—the boss is dead."

"Like hell he is," came Pulaski's response. He lay in the ash, his hands and arms burned and one eye blinded from the smoke. The Big Burn, as the fire came to be known, gutted Wallace and killed at least eighty-seven firefighters and townsfolk. Five of those men were under Pulaski's command. He never fully recovered. For the rest of his life, he tended to the graves of the fallen and rarely left his home and blacksmith shop in Wallace, where he invented his namesake tool and the icon of wildland firefighting. The ax on one side of the tool's head can be used to chop limbs from trees or cut thick roots; the adze on the other end can clear pine needles and leaves from the forest floor. As such, from Colorado's plains to Southern California's brush to Tennessee's thick timber, the Pulaski works well in every fuel type. Today, firefighters' caches all across the country store tens of thousands of the tools that bear Pulaski's name.

At least four of the hotshots on Granite Mountain were using Pulaskis the day of the training fire. Donut, who is six feet tall, had customized his with a longer handle. As lead Pulaski, he went first, cutting roots and mats of pine needles and otherwise setting the path for the line. The ten other hotshots in the scrape followed behind. A few feet separated each man, and, unlike out ahead where the saw teams were working, it was relatively quiet in the scrape. Donut and the veterans in the scrape talked about baseball, beer, girls, and food as they swung their tools.

Swinging hand tools at a rabbit's pace doesn't generate the same risks as running chainsaw or swamping, but it comes with its own special brand of hazards. There's the obvious—the business end of the tool. In the few hours they'd been cutting line, Donut had already taken a few thousand tool strokes, each time flinging an adze full of dirt into a berm of cut brush, limbs, and pine needles forming on the green side of the line. Already his pants were covered in dirt, and dust filled the creases on his face. At one point, the Pulaski's head skipped off a rock and clanked into Donut's shinbone. He let out a stream of curse words and felt a wave of nausea. The hotshots around him winced, but they'd all felt the same thing.

Donut's momentary pain didn't slow the otherwise steady pace of the work for long. What did is when Donut attacked a forearm-thick root that crossed from the black (or future black) to the green. If the root burned, the prescribed fire could spread across the line and escape. When Donut paused to chop through it, the other hotshots stacked up behind him and progress slowed. But Donut stubbornly refused to admit defeat to a root. He switched to the Pulaski's ax end, wound up into an overhead swing, and buried the sharpened bit into the root. Still, it didn't break free, and from the back of the line came the call "Take a bump!" It meant Donut needed to leave the root for the next guy and keep advancing the line.

The call had come from Bob Caldwell, the squad boss working at the back of the scrape. After five seasons fighting fires, Bob had proven himself a strong leader and over the winter of 2013 had earned one of the three coveted squad-boss positions on Granite Mountain. At just twenty-three, Bob was the youngest, least experienced, and baldest of them. In all likelihood, he was also the smartest. He loved Hemingway, Coors beer, and hotshotting; on a trellis in his backyard, he'd hung a wooden sign that read, I'D RATHER DIE IN MY BOOTS THAN LIVE IN A SUIT.

In addition to keeping the scrape's pace steady, Bob's job during the training day was quality control. With a tiny hand rake called a monkey paw, he swept from the line whatever the others left. He checked the dirt berm for flammable material. He ordered sawyers to fell dead trees that could burn through and fall across the line and had hotshots with Pulaskis chop back pockets of light brush that could kindle a fire. He paid attention to the line's width and the depth of trenches dug below steep slopes to keep burning pinecones from skipping across the line and igniting the unburned forest below the men. There was a certain art to deciding where trenches were needed. Sometimes three feet deep, with berms of the same height, trenches took extra time and effort to build and slowed line production. Build too many and the crew would be too slow to catch the fire; build too few and you'd jeopardize the work already put in.

"Pulaski and two rhinos back!" Bob yelled forward more than once that day when he wanted the men to deepen a trench or widen a piece of line.

It was now, before the fire season really began, that the rookies and new guys needed to learn the crew's standards. The line is the mark that hotshot crews use to judge one another, and Marsh, Steed, and now Bob all insisted on perfection.

The crew had settled into a pace after a few hours of cutting line. Up front, the chainsaws whined steadily and the men swinging hand tools worked amid the syncopated rhythm of metal slamming into dirt. All day, Steed had been moving between the head of the flag fire, to scout, and the back, to rejoin the crew. When he returned, he'd dive in to help the scrape or the swampers, and his very presence drove the men to work harder. But that afternoon, Steed stepped up the challenge. He called the squad bosses on the radio.

Something was up. Steed mentioned a spot fire, a small blaze ignited by sparks thrown out ahead of the main fire. The occurrence of spot fires is a classic warning sign that a fire is becoming increasingly active and dangerous. Moments later, the radio crackled again: Steed couldn't handle the spot alone.

The saws fell quiet and all the hotshots hustled back to where Steed had already started cutting line around a new series of pink flags hung in the woods. If they didn't catch it, the spot might become slop-over—when a fire jumps a piece of line and takes off running through the forest. If that happened, Granite Mountain's work would be for nothing, and the men would be exposed to flames. The pace approached a sprint. Scott stood in a shower of wood chips as he and the sawyers redlined their saws and tore a path through a pine stand. Behind them, Joe and the swampers grabbed cut brush, saplings, and branches by the armful and threw the vegetation over the edge of the line. Bob pushed the scrape to move faster still. "Hit and lick!" he screamed. The scrape followed the order, stepping forward after each frenzied tool stroke.

The crew couldn't move fast enough. Steed announced more new fires every few minutes. Another spot fire appeared out ahead of them. Then still another. Catching the spots was never the point of the drill. Granite Mountain needed to escape.

Before they left the station that morning, each hotshot had equipped himself with a training fire shelter, a green plastic version of the aluminum heat shield that every firefighter in the country carries on the line.

Weathering flames beneath the cover of a heavy blanket isn't a new concept. During the Corps of Discovery's trip westward in 1804, Captain William Clark noted the use of very basic fire shelters.

"The prairie got on fire and went with such violence and speed as to catch a man and woman and burn them to death. Several escaped. Among [them was] a small boy who was saved by getting under a green Buffalo skin. They say the grass was not burnt where the boy sat," Clark wrote in his journal.

Today's shelters are effectively aluminum pup tents that act as one-time-use heat shields. They're not designed to withstand direct contact with flames—they melt—but lying beneath them, firefighters have survived entrapments even when temperatures outside the tents reached into the thousands of degrees Fahrenheit and temperatures inside climbed above three hundred degrees. Each shelter weighs just four pounds and is carried in a bread-loaf-size pouch on the underside of a backpack. Since the late seventies, when the Forest Service mandated that all of its wildland firefighters carry such shelters, they've been deployed more than 1,230 times. Between 1926, when reporting statistics became reliable, and 2012, 392 American wildland firefighters have been burned to death. Without shelters, that number would likely be much higher.

Despite evidence of their effectiveness, shelter use isn't universal. In Australia and Canada, countries that also vigorously fight wildfires, firefighters don't carry shelters, and fatality rates are comparable to those in the United States. Australia's and Canada's logic is that simply having a last-ditch survival mechanism increases the likelihood that firefighters will make the types of risky decisions that lead them to depend on an aluminum tent to survive. In other words, they believe carrying a shelter makes firefighters gamblers. Some hotshots in the United States subscribe to this belief, but the vast majority accept the extra weight as a reasonable, though unreliable, safety net that they hope never to use.

Granite Mountain moved together in a line as Steed led the hotshots through the pillars of pines. A few men broke into a jog just to keep up. Occasionally Steed glanced back, as if the fire were behind them and getting closer, and the farther they hiked, the faster they moved. Steed's intention was to teach the hotshots how to cut through the drowning hum of their heartbeats and default to their training. Finally, he ordered them to drop their packs.

"Grab your shelters," Steed yelled.

Grant, toward the back of the line, undid his waist buckles, grabbed a bottle of water, and ripped the fire shelter from the bottom of his pack. The pause took no more than thirty seconds, and the men left their gear scattered throughout the forest. The single-file line dissolved into a loose pack, and they continued to flee through the forest. A few tripped on logs and bushes, pushing themselves up with gloved hands, while the crew's three squad bosses howled behind them: "Go! Go! Go!"

Steed pushed the crew for longer than was comfortable, then longer still. Finally he gave the order: "Deploy!"

With a few hurried tool strokes, Grant cleared away the pine needles that could ignite below him in a real fire. He flicked open the practice shelter as if spreading a quilt over a bed, pulled the tarp over his head, and fell face-first into the cool dirt, with his boot heels toward the flag fire and his gloved hands pinning the shelter to the ground. He dug a small hole in the dirt to bury his nose. Cool air sinks. Sunlight filtered through the green plastic training shelter, and he waited in the green glow and the smell of dust.

Grant heard the rustling of other hotshots settling into their places, the slowing of his heartbeat, footsteps drawing closer. Suddenly, somebody grabbed the peak of his tent and jerked it—up, down, left, right—to simulate the hurricane-force winds, sometimes as high as sixty miles per hour, that accompany fast-moving flames. Grant pushed his weight into the tarp and held on, lowering his head as the plastic beat against his helmet. Firefighters have survived under their shelters for more than an hour. Steed let his crew lie in the dry heat for a fraction of that, but a few minutes was plenty of time for Grant to contemplate the horrors of a fire entrapment.

CHAPTER 3

MARSH'S CREW

By the time Granite Mountain finished its deployment exercise, the Arizona sky glowed with the brilliant pastel colors of a Southwest sunset. For the hotshots, the flag fire was over, but even after a long day, Steed had no intention of capping off the training event of the season with a pretty sight. The men lined out behind their captain, and once again Steed took off up the hill.

Renan Packer, a quiet and solidly built rookie who sat across the aisle from Grant in Alpha's buggy, was wearing down. After cutting line all day, his tool felt as heavy as his mouth felt dry, and every step taxed muscles long since drained of energy. He hiked so close to the man in front of him that when the pace slowed on the steepest pitches, he slammed his helmet into the underside of his crewmate's pack.

Renan glanced uphill, where he hoped to see the terrain flatten or, even better, see the buggies. He saw neither, nor did he see Marsh approaching. The crew's superintendent had moved the buggies to camp and walked down the trail a few hundred yards to watch the men hike in. For the longtime superintendent, twenty men marching together as a single unit was a beautiful sight. But a glitch marred the symmetry. Renan had tucked the handle of his tool into the back of his pack to ease the burden on his exhausted arms. The unconventional approach broke with procedure, and even if he wasn't acting superintendent, Marsh wouldn't let that fly on Granite Mountain.

"Fix it, Renan," Marsh said. "Carry that fucking tool the way you're supposed to!"

Marsh's scolding snapped Renan from his exhaustion-induced delirium, and he dropped the rhino to his flank, where it was supposed to be. He put his eyes straight ahead and tried to shut out the torrent of expletives still pouring from his boss's mouth. The veteran hotshots told stories of Marsh bawling out new guys for sticking the tips of their running chainsaws into the dirt, falling out of hikes, or misusing a Pulaski. One said that the first time Marsh spoke to him, it was with his middle finger raised—a joke, but one the rookie found intimidating. The ass chewings their superintendent delivered and his general surliness were the stuff of legend. Regrettably, Renan now understood what the veterans were talking about.

Superintendents shape their crew's personality. A superintendent's relationship with his crew is like that of a coach to his athletes, but Marsh held his men even closer. He called the hotshots his kids—they called him Papa—and Marsh joked with his wife, Amanda, that they didn't need children because they had nineteen already.

Marsh wasn't a westerner by birth, but the West had always appealed to him. He grew up the only child of a biology teacher in a small town in North Carolina's Appalachian Mountains. As a kid, he was outdoorsy and used to steal bareback rides on his neighbor's horses, but in high school, Marsh temporarily abandoned his outdoorsy passions, trading his western garb for Polo and earning the nickname Biff, as a reflection of his preppy dress. After high school, he spent a few years dabbling in a wide range of pursuits. He worked construction, learned to weld, and was an aspiring horseman, a rock climber, and an avid mountain biker. He learned to shape leather into saddles, fixed bikes, and painted logos on ambulances. He studied biology at Appalachian State University. He married young.

Marsh took his first fire job—with an engine crew in Arizona's Tonto National Forest—while still in college. At season's end he returned to North Carolina to finish his degree, which he completed in 1992. Following a divorce, he left the South and started over in Arizona. He quit fighting fire for five years and bounced between jobs throughout the nineties. During his hiatus, Marsh took classes in practical skills—Microsoft Office Applications—and got by with odd jobs that relied on his ability to work with his hands.

In 1998, he returned to work for the Forest Service as a firefighter. Marsh was twenty-eight then and the agency hired him as the captain of a two-to-five-person engine out of Globe, Arizona. In that remote and dusty mining town, Marsh felt the allure of hotshotting. When the Forest Service's Globe Hotshots needed an extra hand, he occasionally filled in as a squad boss in charge of as many as ten men. Though Marsh never worked full-time as a member of Globe's overhead, or even spent a full season as a crew member, the fires he fought with the hotshots left a lasting impression. The speed and quality of line construction, the rookie intimidation, the brooding way the superintendent distanced himself from the crew—many elements of Globe's firefighting style would resurface on Granite Mountain when Marsh became the crew's superintendent almost a decade later.

Despite his attraction to the job, Marsh took five more years before fully committing himself to wildland firefighting. He was married again. He'd met his second wife, Kori Kirkpatrick, in the early nineties while battling a blaze in Texas. Kirkpatrick was also a firefighter. Driven and cerebral, by 2000 she had begun making plans to found a fire academy. Marsh, apparently in need of change, left his engine captain's position in Globe after just two fire seasons, and the couple moved northwest to the cooler air of Prescott. For the next few years, as Kirkpatrick got the fire academy up and running, Marsh pieced together jobs, working as an emergency medical technician on a local ambulance, as a supervisor for a small industrial company that employed a fire engine for safety measures, and eventually, for a few short months, as an instructor at his wife's fire academy. But by the spring of 2003, Marsh was thirty-three and tiring of jobs that rarely exceeded

$12 per hour. He applied for a career-oriented, full-time position with the City of Prescott's new Wildland Division, which specialized in proactively preparing the city for wildfires.

In the years before, California's Hunt Research Corporation had named the Prescott area one of the ten places in America most likely to be hit by a wildfire. Because of this humbling reality, the city formed Crew 7 for the sole purpose of creating defensible space, or thinning the thick ponderosa forests and brush around town. The area of thinned vegetation was designed to prevent fires that started on the outskirts of town from burning buildings. Depending on budget, crew size oscillated anywhere between three and twenty members. Initially, Prescott hired Marsh to be a squad boss, and he quickly moved up to be second-in-command on Crew 7.

The work required all the sweat and labor of a firefight but offered little of the challenge and none of the pride. For Marsh, who learned to love the firefight while working with the Globe Hotshots, that posed a problem. Soon after becoming the captain, he made it a personal goal to turn Crew 7 into professional firefighters. He had a long way to go.

Most jobs on Crew 7 paid only slightly better than minimum wage, and because of this, it didn't attract career-driven firefighters. The city outfitted the men with fire-resistant clothes torn from years of use and relegated the crew to a barn situated amid a formation of granite domes, by an almost toxic lake. Not only was the city dump across the way, but in the 1980s, a radiation test at a well house not far from the station had detected radon—a gas emitted by the unusually high uranium content naturally found in the granite—at the highest levels ever measured indoors.

"We'd leave to go for a run and the fucking rats would eat our lunches," said Phillip Maldonado, a hotshot who was with the crew from the beginning, until early 2013. The city's structural firefighters didn't welcome Crew 7, either. They couldn't understand why a bunch of feral wildland guys were allowed to wear the proud emblem of the state's oldest fire department. Once, at a department Christmas party, one of the structural firefighters walked up to Maldonado and asked, "Who the hell invited you?"

As captain, Marsh was heir apparent to the downtrodden crew. He regularly fought with the city on the men's behalf, arguing that they needed more money for equipment and gear, and higher pay. But his troubles were bigger than Crew 7's hard luck. His second marriage, to Kirkpatrick, was falling apart, and though he couldn't hide the hurt—his colleagues considered him a moody guy—neither could he bring himself to talk about it. Once, halfway into a twenty-hour road trip to Idaho, Marsh told the only other guy in the truck, "I'm getting a divorce." Then he didn't speak for the rest of the trip.

As his personal life unraveled, Marsh became hard to be around and discontented with his job. In one six-month performance review in 2005, he wrote that by the next year, "I would like to be the superintendent of Crew 7." His supervisors rebuffed the request. For all of his competence as a teacher and organizer, and all his skill as a firefighter—"he knows what to do on the line in wildland fire situations," his supervisor wrote in 2005—he had shortcomings as a leader and seemed unable to separate work from his private life. He had a tendency to distrust his subordinates and micromanaged. One review said, "He needs to learn not to yell when jobs are not performed to his standards . . . he becomes very agitated and does not accept positive criticism well. Blames other people for his misgivings." His score for interpersonal relations was "Does Not Meet Expectations."

Marsh's situation started improving toward the end of 2006. He'd finalized his second divorce. Crew 7's superintendent left, and Marsh's supervisors had decided that his job performance had improved enough for him to take over. The year before he became superintendent, Crew 7 had earned the status of Type 2 Initial Attack Handcrew, which is a certification one tier below hotshot. They were now qualified to fight fires anywhere in the country, but that didn't guarantee they'd get many opportunities to travel. If the fire risk was high around

Prescott, Crew 7 would assuredly be kept behind to guard the home front. Hotshots, though, are a nationally controlled resource available to fight the country's highest-priority blazes regardless of the fire situation at home. Marsh craved for Crew 7 the same freedom and prestige he'd seen on the Globe Hotshots. But he couldn't get it without the city's approval.

Few people in Prescott saw the point, though. If Crew 7 were to become hotshots, they'd be spending less time thinning brush and more time on the front lines of the nation's biggest fires. City officials, whose approval he needed to start the project, preferred to keep Crew 7 at home to protect Prescott, not the hundreds of other western towns threatened by fires every summer. Let the feds and their bottomless budgets do that.

Still, against fierce opposition, Marsh persisted. In meetings with his chiefs and city councillors, he argued that hosting a hotshot crew would lend more credibility to Prescott's already esteemed fire department. Prescott's department, the oldest in Arizona, dated back to the days of the Earps and counted among its alumni a governor. But there were also financial incentives for making Crew 7 into hotshots. The Wildland Division was funded primarily through grants given by state and federal agencies to towns that proactively prepared for wildfires. Provided they kept creating defensible space around Prescott, Crew 7 could still be funded by grants. But when they went to fire assignments, the city would reap the financial rewards. Every time the men worked a wildfire, the state or federal government—whoever funded the firefight—would pay the city the equivalent of low-cost rent for the crew: $39 per hour, including the men's wages, but not fuel, equipment, or lodging when they needed it. By allowing Crew 7 to become hotshots, Prescott could continue preparing for future wildfires while at the same time offsetting the cost of hosting a fuels crew, reaping their benefits, and breaking even on the deal.

After years of battling bureaucracy and politics, Marsh and the fire department finally swayed the city council to approve his request to pursue Crew 7's hotshot status. Now Marsh faced the much larger task of actually building a hotshot crew, a process that can take decades. First, he needed a roster that included at least six dedicated, experienced, and professional wildland firefighters—a good overhead. He'd also need a medic, a crew member capable of loading twenty firefighters and their gear into a helicopter, a pair of squad bosses to oversee nine crew members, and a captain qualified to step into the superintendent position if Marsh was injured or worse.

Marsh was hamstrung from the start. Prescott's fire department paid its Wildland Division a dollar less per hour—$12 for most rookies—than the federal wildland crews did. With twenty other hotshot crews in the Southwest to work for, this limited Marsh's hiring prospects. A few times, Marsh took a risk on a candidate—hiring former inmates and recovering drug users—only to have him or her relapse in the middle of the season.

But for all the odds stacked against him, Marsh had one thing to offer that no other hotshot crew in the country could: access to the city's better-paying and more stable jobs on the red trucks. The Prescott Fire Department permitted the men who proved themselves on Crew 7 to ride along with the city engines, which required more education and experience than an entry-level wildland job. Over the years, the structural department hired thirteen Crew 7 alumni. Marsh made the most of his competitive advantage. His staff became progressively stronger as he poached men from other hotshot crews looking to make the leap to structural firefighting, and he took risks on young applicants with no fire experience at all.

Some of his best firefighters came to him just out of high school, promising that as athletes—linebackers, bull riders, wrestlers, and runners—they'd already proven that they had the toughness required of hotshots. Marsh recognized that their sports training was a good physical foundation but far from a guarantee that they could handle the marathon of a fire season. When he hired, he told the rookies, "You're no longer an individual. Whatever you do affects twenty other guys."

If the new hires could handle it, Marsh shaped them into firefighters with a few months of hard work. He hammered into them the fundamentals: how to swing a Pulaski, run a chainsaw, spark a backfire, and work through the pain. Marsh had helped with hiring and training long before he became superintendent, and by 2006, when Crew 7 attained Type 2 status, he was supported by eight or nine men who believed fully in his vision. Still, whenever the crew was sent to a fire, Marsh was once again forced to prove himself—this time to other crews.

"We were the redheaded stepchildren," Maldonado said. "They didn't want us at home, and they didn't want us on the fire line."

For one thing, Crew 7 looked different from other fire crews. Instead of buggies, those burly custom people movers, Crew 7 drove white twelve-passenger vans. The rigs weren't capable of handling the wear and tear from twenty young men with fierce addictions to chewing tobacco and little allegiance to sanitation. If buggies get nasty— they do—then the vans were leagues worse. The fire gear was stored in the back, and the seats were so full of dirt that clouds of dust would puff up when the men sat down. The air-conditioning in one van didn't work at all. As a joke, the unlucky squad assigned to the hot van tucked canned tuna fish into the functioning cooling vents of the other van—the smell was putrid. When Crew 7 rattled into fire camp, the other hotshot and handcrews immediately took notice. Most cringed.

"Hotshotting is a good old boys' club, and new crews are never welcomed, especially when they're from other agencies. Whatever Marsh went through wasn't unique," said Jim Cook, who was the superintendent of the Arrowhead Hotshots, out of Kings Canyon National Park, in the 1980s and 1990s. Arrowhead was among the country's first Department of the Interior crews. "But the shit you get as an upstart isn't for performance. It's for tipping the world sideways."

When Cook started Arrowhead, the established hotshot superintendents were far from welcoming toward new crews. Back then, many crews still had ties back to the hotshots' predecessors, the Civilian Conservation Corps fire crews and the forty-man fire teams that sprang up in Oregon in the late thirties. Though these teams were effective, research on line-construction speed, logistical ease, and the simple management challenge of wrangling forty young men prompted the crew size to be cut to twenty firefighters. The first hotshot crews—the name referred to the fact that they were always given the most intense assignments—emerged in their present form in the late 1940s, after many years of wildfires threatened thousands of homes in Southern California. By the 1960s, the Forest Service, the agency primarily responsible for the creation of hotshots, had stationed nineteen crews near major western airports so they could be flown anywhere in the country within twelve hours. The maturation of President Eisenhower's Interstate Highway System nullified the Forest Service's need to fly crews to fires. Good roads made it cheaper and more efficient to drive the men to blazes, and today hotshots rarely leave their crew buggies behind.

By the nineties, hotshots had evolved into an esteemed organization in the wildland fire community, and the crews became accelerated pipelines for career-driven firefighters. In a matter of a few large fire seasons, a hotshot could spend more than 250 shifts on fires—more than most structural firefighters spend on blazes over their entire careers. Because of their ample fire-line experience, a select few hotshots go on to highly skilled positions in fire management, becoming smoke jumpers, incident commanders, investigators of fire fatalities, or, like Cook, academics whose research can shape the future of fire-line decision-making.

Though the early hotshots were much more militarized and in vastly better condition than the civilian teams Forest Service rangers like Ed Pulaski had raised to battle violent blazes decades earlier, they still relied on only a few qualified firefighters to make decisions. The biggest difference between the new crews and the old militias came down to cultivated pride. The young hotshots believed they were the best firefighting force ever created, and that anybody who cared to claim the title needed to earn it first.

While Crew 7 were trying to become hotshots, during one shift on a blaze in Oregon, a firefighter from another crew grabbed a can of spray paint and drew a line in the middle of the road. Next to it he wrote, DON'T CROSS IT. Many firefighters simply refused to talk to the guys or eyed them up in the chow line. The Forest Service firefighters' concern, whether justified or, much more likely, not, was that a municipal crew wouldn't have their backs on the fire line. Marsh and his men were officially given a chance to prove themselves in 2007, when Crew 7 was made a hotshot training crew—one step beneath actual certification.

By then he'd changed Crew 7's name to Granite Mountain and upgraded its second-rate gear to the best equipment available. A sign in their gear cache read, TOTAL COST OF A WELL EQUIPPED HOTSHOT: $4,000. Marsh sold the vans and bought two $150,000 white-and-red crew hauls that, when parked beside the green trucks of the Forest Service and the yellow trucks of the Bureau of Land Management, announced that Granite Mountain was proudly different. Perhaps best of all, Marsh moved the men out of the rat-infested radiation barn by the lake and into the downtown fire station. The city gave Marsh $15,000 to fix it up. To get the most out of the money, he and the crew did much of the work themselves and cut costs wherever possible, even scrounging chairs from the side of the road or the dump. They had Steed's brother install the linoleum floors, inlaying in the white floor black tiles that read GMIHC (Granite Mountain Interagency Hotshot Crew). If any rookie touched the black tiles, he had to do ten push-ups.

Granite Mountain's two seasons as a hotshot-crew-in-training were intense. It started with a live certification drill, similar to the one 2013's crew would be going through in Prescott National Forest but with considerably more scrutiny. A pair of longtime Forest Service hotshot superintendents shadowed Marsh and his firefighters as they were put through a series of tests meant to replicate any scenario hotshots might encounter on the fire line. The other superintendents threw at Marsh and his crew a drill that demanded they guide in air tankers, coordinate fire-line operations with bulldozers and a half-dozen fire engines, operate without Marsh as superintendent, and decide when it was safe to attack a blaze and when they needed to step back and watch the fire burn. The crew put out rapidly expanding spot fires, felled enormous trees, and dealt with medical emergencies in the form of bee stings, heat exhaustion, chainsaw cuts, and burns.

Marsh and Granite Mountain excelled, but they wouldn't hear the other superintendents' ruling for months. In the meantime, they returned to the fire line, where the tests kept coming. They stopped one fire from scorching a subdivision near Payson, on the notoriously dangerous Mogollon Rim, and another by accepting an assignment that no other crew on scene would take: a thankless and grueling job that required cutting twenty-six hundred feet of line from the rim of Idaho's Hells Canyon to the Snake River. Despite these accomplishments, it was August 2008 before the crew heard back about the ruling on Granite Mountain's hotshot status. At that point they were working a fire in Northern California's Klamath National Forest, a swath of 1.7 million acres of ponderosa pine, Douglas fir, and poison oak forest that drops from nine thousand feet to the Pacific Ocean in less than fifty miles. That morning, the smoke hung in long white strips above the many river valleys cutting through the peaks.

Marsh gathered the men beside his new Dodge 4500 with its customized taillights. Next to the initials was a big *T,* for "training crew," which Marsh had taped there to ensure other hotshots knew that Granite Mountain wasn't claiming a title it hadn't yet earned.

"I got the call," Marsh told them. He flicked out his knife and scraped the *T* from the back of his truck.

"We're hotshots," Marsh said, allowing himself a smile. Granite Mountain was one of the only municipal hotshot crews in the country. "This honor is yours, gentlemen. You earned it. Congratulations." Then he led his men back out to the fire line.

Back in Prescott during the tail end of 2013's training drill, the single-file line of panting hotshots broke up at the top of the dusty trail in the Prescott National Forest. Renan, still gasping, followed the other hotshots in Alpha squad to their buggies. Somebody put on a Tupac album, and the heavy beats of gangster rap thumped across the campsite. As the men refilled their waters and chainsaw gas, Marsh watched, as he'd done all day, how the hotshots performed. He didn't camp with the men that night. Not long after sunset, Marsh drove away. It was the last time many of the hotshots would see Marsh until he rejoined the crew near the end of June.

Renan and Grant climbed into the Alpha buggy. Clayton Whitted, Alpha's squad boss, had ordered the two to keep track of each other throughout the season. They'd help each other run errands in fire camps, keep track of the other guy on the line, and even be there when the other went to the bathroom. Grant, amused by the idea of having a minder, called Renan his "battle buddy." Cheesy and flippant, the catchphrase made a few of the senior guys laugh, but it irked others.

As rookies, Clayton had given the pair one more lowly job: unloading the other hotshots' overnight gear. Along the ceiling of each buggy ran a shelf that held the crew's black duffel bags, each filled with a sleeping kit, a few changes of clothes, a dozen clean socks, and various creature comforts—books, journals, men's magazines, portable music players—to entertain them during moments of calm on the line.

"Brutal day, huh?" Renan said, tossing a bag to Grant. "Eric was riding me pretty hard back there."

"Yeah, that sucked," said Grant, who caught the bag and threw it into the pile outside the truck. "Did you hear Scott tearing me a new one?"

"Hilarious, man. I saw you go down and thought for sure you weren't coming back up," Renan said. "I was like, well, looks like I'll have a new buggy mate. Grant's knocked out."

"Rookie!" one of the veterans called into the truck. "Toss me a Gato."

Renan opened the cooler and underhanded a cold bottle of Gatorade to the veteran, who nodded in thanks.

"Nah, I just got too hot," Grant said. "I had my yellow buttoned up to my neck." Renan laughed at the thought of his overdressed friend sweating profusely and overheating from a wholly preventable cause.

"Puking always makes me feel better anyways," Grant said.

Meanwhile, in Bravo's buggy, Brandon Bunch, a former bull rider and fourth-year sawyer, and Wade Parker, his swamper, had pulled the type of move hotshots would talk about for years to come. That morning before their shift, Wade had bought twenty steaks, and Bunch had thrown a propane grill beneath his seat in the buggy. He knew the hotshots always camped after their training day, and that meant Meals Ready to Eat, the pre-packaged military breakfasts, lunches, and dinners that crews eat on fires. The veterans didn't need practice eating MREs.

MREs have been a staple of the hotshot diet since they were developed, during the Vietnam War. Over the years, the variety and quality of the food has come a long way from "Ham Slice" and "Frankfurter and Beans." Wars tend to increase the variety of MREs in circulation, and since 1993, more than 241 new items have been approved. Some, like Chicken Pesto Pasta with Patriotic Cookies, are decent. But the meals, which are heated by a water-triggered chemical reaction, always produce a nauseating stench. Granite Mountain would have plenty of opportunities to compare tasting notes on the variety of MRE flavors. Bunch and Wade knew as much. Steaks were a rare gift—one they were happy to give.

"For everybody, but the rookies gotta eat their MREs first," Bunch said, sparking up the propane grill. "Sorry, fellas." He happily ensured that this year's rookies went through the same traditions he had: The veterans picked the rookies' meals.

Anthony Rose, a second-year hotshot originally from Illinois, ripped into the boxes and started sorting through the flavors. He'd come to Granite Mountain via a firefighting position on an engine in Crown King, a 164-person community east of Prescott. He'd been picked on regularly in his rookie year. The memory of the hazing he'd weathered was still fresh, maybe even magnified by his own marginal position of power.

"Dig in, Grant," Tony said as he tossed him an MRE.

After a long day of swinging a tool, hunger usually makes it possible to forget that MREs are little miracles of science more than they are food. But that night, Grant couldn't get excited about eighteen-

month-old, deoxygenated meat when the men he'd worked with
shoulder-to-shoulder all day were celebrating the end of their training
with steak.

Grant took up a spot by the campfire, where he tore into the pack-
age and chewed through powdery biscuits while Tony, Bunch, and the
veterans laughed over the grilling meat.

Grant was used to being treated with a certain level of respect. He
could agree to play the hotshots' reindeer games while on the crew,
but only if it gained him entry into the club. A charismatic kid who'd
been an athlete in high school, Grant was the well-dressed and earnest
one who moved with ease between all social groups. He didn't hold
grudges and didn't see a reason for social hierarchy. When he'd first
moved to Prescott, he used to join his aunt Linda and a group of her
friends on morning walks, on which he was the only man and the
youngest person by twenty years. Not surprisingly, the women loved it
when Grant joined them, but his charms fell short on Granite Mountain. None of the hotshots cared how somebody was used to being
treated. Respect had to be earned on the fire line. That took time.

CHAPTER 4

FIRST FIRE

After the crew passed its two-week critical, Granite Mountain was available to fight fires anywhere in the country. At a moment's notice, even if rain was dousing Prescott, the hotshots could be loaded into an airplane and shipped to a fire fifteen hundred miles away. Trouble was, few fires were burning anywhere in the country, and that meant Granite Mountain stayed put in Prescott. By early May, the crew had spent weeks working in and around the station, and stagnation wasn't good for anybody. The men wanted to do the job they'd been training to do, and the longer it took to get an assignment, the more likely it became that some bored hotshot would do something stupid and create problems for himself and the department.

Every morning, Steed started the crew's day with the same routine. They met in the ready room, where one hotshot added daily Internet-found factoids to a whiteboard—random things about the amount of milk cows produce, or the hottest days in history—and Steed volunteered another to read aloud the situation report, a summary of all the nation's fire activity. On May 1, activity was light, with fewer than a thousand new starts and fewer than ten of those designated as large, meaning one hundred acres or more if the fire was in timber and three hundred acres or more if it was burning in grass. Given that the Southwest was as dry as in any year in recent memory, it seemed strange that no fires were burning in the region. The last measured rainfall at Prescott's Sundog Weather Station was .11 inch, three weeks earlier. April had been the third month in a row with less than a quarter of Prescott's average precipitation, and Arizona's scant winter snowpack had already melted off of even the twelve-thousand-foot peaks to the north. Whether it was a wayward spark or the men's own restlessness, something was bound to break the calm.

Steed's best tool for taming the impulses of nineteen young men was exhaustion. His morning workouts only got harder as the men's fitness improved, and after a couple of hours of running, hiking, sit-ups, pull-ups, and push-ups, he sent them to thin the forest and brush in J. S. Acker Memorial Park, a city park so overgrown that the police often busted meth addicts getting high in its thickets. Day in and day out, the men ground through the same routine: Steed's training followed by five or six hours of running chainsaws and chipping brush.

The men invented ways to stave off the monotony. Alpha and Bravo competed in anything that could be made a competition: Ultimate Frisbee, cake eating, saw work. Then there were the shenanigans. The squad bosses, Clayton Whitted and Bob Caldwell, both balding, shaved their heads at work and forced Grant to do the same. He shrugged when he showed Leah the buzz cut. Thus far, he'd protested most hazing because he found the rituals childish and a little insulting. This one he couldn't avoid, though: Bob, his cousin, was involved. He had to support him. "Sometimes you just have to give in," Grant said to Leah.

He gave in again on his twenty-first birthday.

"Don't tell anybody," Renan had warned Grant. "Everybody's going to give you shit if they find out."

But Grant either didn't care, didn't believe him, or actually wanted to be the center of attention, because that morning he let the news slip. When Steed heard, he concocted a birthday cake from MREs: crackers, pound cake, and cookies frosted with packets of peanut butter and chocolate-hazelnut butter churned together. The crew sang

"Happy Birthday" and laughed as Grant ate a barely digestible few thousand calories of sugar before the morning workout. He couldn't finish, and Renan chipped in to help, but Grant still vomited up his birthday cake during that day's training.

The calm finally broke on a warm and breezy evening in early May. Outside, plastic bags and wrappers from the food stand across the street pinwheeled into the chain-link fence, where they caught and flapped in the breeze. It was nearly time to go home for the day. Some of the hotshots were sharpening tools or putting an edge on their chainsaw chains, but most of the crew lounged about the saw shop. They perched on the workbenches, snacking on the cookies and crackers taken from MREs and arguing the finer points of hoppy vs. cheap beer. As usual, Renan didn't add much to the clamor. From the first day of the season, his goal had been to earn the Rookie of the Year award, an annual honor that would get his name on a plaque beside Wade Parker and Andrew Ashcraft, two hotshots on track to securing permanent fire jobs—exactly what Renan wanted. To avoid getting singled out, he did the opposite of Grant and kept his head down and his mouth shut.

Renan listened to the endless drone of radio transmissions coming through a speaker set up in the station: a city engine returning to quarters, a battalion chief going off duty for the day, a car wreck. He usually ignored the radio chatter, but today Renan was listening more intently. That morning, during the daily briefing in the ready room, Steed and the crew had talked about the Red Flag Warning the National Weather Service had issued for the Prescott area. It meant that that day the region had entered a period when dry fuels, warm temperatures, and strong winds were aligning. In such conditions, wildfires threaten to be explosive and extremely difficult to control.

Renan felt a surge of adrenaline when he heard radio traffic about a wildfire burning in grasslands near a subdivision twenty-five miles north of the station. Granite Mountain was a half-hour away. Dispatch might send them.

He ran across the pavement to the buggies and pulled his line gear off the shelf and into a pile on his seat. He checked every pocket of the bag to be sure he had everything he needed: extra AA batteries for his headlamps and radio, six quarts of water, an MRE, a warm layer, a Clif Bar, chewing tobacco. He'd been waiting for this moment for a long time.

Some men and women seem born to be firefighters. They're woodsmen from places where the trucks are lifted and the high school parties are held around bonfires on ridges with views of towns like Truckee, California; Missoula, Montana; and Redmond, Oregon. Prescott is such a town. Like nearly half of the crew, Renan grew up there, but none of the guys he went to high school with ever expected that he'd become a hotshot, even if he was a gifted athlete.

At sixteen, Renan contracted the West Nile virus and developed neurological complications related to the pathogen. The symptoms— deafness, paralysis, and seizures—left him in a wheelchair for two months during his freshman year of high school. By his senior year, thanks to a combination of physical therapy, a cocktail of drugs, time, and learning how to better manage the stress that often triggered the seizures, he'd brought the condition under control enough to play catcher on Prescott High's varsity baseball team.

After high school, he wanted to work with his hands. He set his mind to becoming a structural firefighter. He craved the feeling of accomplishment that physical labor provided. He took fire science classes at Yavapai College. But there are far more candidates than jobs in structural firefighting, which requires more education, earns slightly better pay, and offers more professional stability. Despite applying to a few departments, he couldn't get a position. Like Grant, Scott Norris, and many of the other guys, Renan saw Granite Mountain as a way to distinguish himself from the other applicants. Marsh didn't hire him when he first applied to Granite Mountain in 2011; neither his résumé nor his physical aptitude stood out among the twenty or thirty men and women who applied that year. Renan came back prepared in 2013.

This time, he passed the physical test easily. He wore a shirt and tie to the follow-up interview, a highly unusual formality among hot-shot crews. As a means of communicating the seriousness of becoming a hotshot, Marsh required his applicants to dress up and interviewed applicants in the ready room.

When Renan arrived, Marsh sat at the head of a long Formica table in front of the whiteboards. Flanking him were his squad bosses and lead firefighters—five hotshots in total, all wearing their yellow fire shirts tucked into green pants. They asked Renan to sit alone at a table facing them.

Marsh always asked the same question first: "Tell me how Granite Mountain started."

Renan stammered through the answer: fuels crew to hotshots in six years. Marsh, Steed, and the squad bosses kept firing questions his way: "Have you ever lied at work?" "Where do you see yourself in five years?" "What's the hardest thing you've ever been through?" Renan paused for a moment before answering the last one. He never in-tended to bring up the illness. He hadn't relapsed in years, and the doctors didn't think his fighting fires would pose a danger to him or the crew.

"The two months I spent in a wheelchair," said Renan. He told the whole story: the haunting numbness that crept up his spine and sig-naled a seizure's approach, the exhaustion that lingered for months after, the white void of blindness, the fear that his sight might never return, the struggle to accept what he couldn't control, the decision not to let it hold him back. Marsh, Steed, and the squad bosses lis-tened to Renan intently. Interviews didn't usually turn up such capti-vating stories.

Renan was working as a valet attendant at a Scottsdale golf resort when Marsh called to tell him that he wasn't getting the job. Marsh judged the risk to be too great—too much liability for the city and the crew. That afternoon, Renan drove the hour and a half back to Prescott so he could ask Marsh to his face what it would take to change his mind. He told Marsh that the illness had passed and that if he hired him, he'd do everything to stand out—his career was at stake. A week after that meeting, one of Granite Mountain's top applicants accepted another job. Renan was again parking cars at the golf course when he got another call, this time from Steed. "How would you like to be a part of the crew?" he asked. Renan danced a celebratory jig while Izod-clad retirees milled about the clubhouse.

Back at the station, the door of the saw shop swung open and Steed marched in. "Load up! We got a fire!" he boomed. The men hustled to the rigs, and by the time Grant got in the back of the Alpha buggy, Renan's seat belt was already buckled.

The hotshots caravanned—Steed's supe truck up front, followed by Alpha and Bravo—as they drove across the flats of the Prescott Valley and toward a plume of light gray smoke. Around noon, the grass alongside Perkinsville Road had caught fire, and gusty winds had pushed the flames to nearly a thousand acres.

Donut, who controlled the stereo, put on a rap tune, "Bugatti," by Ace Hood: "Damn, life can switch on you in a matter of seconds." Donut slapped the ceiling with the flat of his hand, yelling out the chorus: "I woke up in a new Bugatti!"

Kevin Woyjeck, a twenty-one-year-old rookie on Bravo squad, thumbed out a text message to his dad, who was a firefighter in Southern California.

WOYJECK: 25 miles away . . . We see smoke.
DAD: Calm . . . Get your calm on.
WOYJECK: Everyone's calm. Don't worry. I'm super. Calm.

Renan peered out the window, hoping to glimpse the fire. A helicopter worked the blaze, and sunlight glinted off the fire engines already on scene. He paid closest attention to the column.

Smoke tells a fire's story; it's the signature. The obvious by-product of combustion, smoke is a mix of evaporated moisture and released gases. Bits of charred wood and soot wafting upward give the visible vapor color. Heavy materials—timber, oil, and houses among them—

that burn hotter and more slowly also tend to produce darker smoke. But on the prairie fire, the smoke was thin and white, angling eastward over the plain with the breeze. It told Renan that the blaze was a wind-driven brush fire: cooler than a timber burn but in quick-burning brush and grasses. Firefighters tell legends of brush fires that, during the windiest days in Southern California, keep pace with interstate traffic. Grass and brush fires are the most volatile burns.

"Yellow up, boys," Clayton, Alpha's squad boss, yelled to the back from his seat in the cab. Renan's mouth was dry. He and the other hotshots put on their fire-resistant shirts.

When they arrived on scene, the fire turned out to barely justify Donut's choice of the "Bugatti" song. A total of thirteen engines, an air tanker, and a hundred-some firefighters had beaten Granite Mountain to the blaze. The blitzkrieg had worked. The fire was mostly out.

While Steed went to get Granite Mountain's assignment from the incident commander, Renan took in the whole scene. The fire's scar looked like burned cropland. Countless wisps of smoke screwed skyward, but only one pocket of active flames remained. The hotshots had shown up too late to catch the large air tanker dropping slurry near the subdivision, but a helicopter still worked overhead. One of the engines on scene was still knocking down the flames with a stream of water.

"Hose!" one of the enginemen yelled.

The call signaled to his crew members that they needed to extend what's called a progressive hose-lay. One of the crew dropped to his knees and shut off the water just behind the nozzle by pinching the hose closed with an industrial-strength clamp. Meanwhile, another man added a hundred-foot length to the end of the lay, reattached the nozzle, and released the clamp. The whole operation took less than a minute, and with water and a hundred more feet of hose to drag along the fire's edge, the crew resumed spraying down the flames. With engines on all flanks of the fire, there wasn't much work left for Granite Mountain. Had they arrived when the blaze started, they may well have witnessed some incredible fire behavior—spot fires, little flaming tornadoes. Instead, the wind died, and so did the flames.

There wasn't much work left for Granite Mountain to do. Steed directed Alpha's and Bravo's scrape to put in a quick line. He sent a few guys to burn out a small pocket of unburned grass inside the fire's interior, thereby limiting the likelihood of spots by controlling the burn's intensity. Once that was done, he and the hotshots "mopped up," or used hand tools, dirt, and water laced with foam to douse the dangerous embers closest to the lines. The shift barely constituted hotshot work.

Donut and the veterans hung around the spots of cooling ash, loud and laughing, as they poked fun at one another and agreed that mopping up, slow as it was, was a lot better than running chainsaws at J. S. Acker Memorial Park. The prairie fire was little more than a welcome smell on their clothes. Not even Renan was terribly impressed. But as the sun slid beneath the horizon, Renan, working amid the twists of light smoke with ash covering his hands, couldn't help feeling pleased.

CHAPTER 5

JUSTIFIABLE RISK

Early one Sunday in May, little Ben and his big brother, Jacob, climbed out of their new race-car-shaped beds and picked their way through the flotsam of Tonka trucks and open books in the hallway. The boys pushed open their parents' bedroom door and crawled into bed with Brandon Bunch and his wife, Janae. Bunch groaned. He was twenty-two, with the dark-haired good looks of a NASCAR driver, but at five-eight and 140 pounds, the fourth-year sawyer was one of the smaller guys on Granite Mountain. He liked sleeping in on his days off. The boys curled into balls against their parents, and for a few short moments the family lay quietly. Then Jacob asked when Garret was coming.

"Not yet, Jakey. He'll be here in a few hours," Bunch muttered with a slight lisp.

Janae, who was seven months pregnant, got out of bed. She knew Jacob was too excited to stay quiet. Garret Zuppiger, one of Bunch's closest friends on the crew, was coming to breakfast before church. Janae led her boys to the kitchen—Jacob first, Ben dragging his blanket behind—put on coffee, popped bread into the toaster, and tuned the TV to the Cartoon Network. Bunch emerged from the bedroom shortly after. Sleeping through the shrieking of *Dora the Explorer* wasn't possible.

One perk of the season's slow start was that the hotshots had more time with their families. Considering the hotshots' youth—their average age was twenty-seven—Granite Mountain was more family-oriented than most crews. Whether they were from Prescott, like nine of the guys, or were recent transplants, most of the hotshots shared small-town values. Many of the men went to church together. Eleven of the hotshots were married, three more were engaged, and nine had kids. Collectively, the hotshots had fathered fourteen children, and by year's end four more, including the Bunches', would come into the world. If the hotshots hadn't already sunk their roots deep in Prescott, many were starting to.

The Bunches owned a two-bedroom home in a new subdivision twenty miles north and east of Prescott and rarely locked their doors. When Zup arrived a few hours later, he walked right in. Jacob sprung up from the TV.

"What's up, buddy?" Zup said as the boy attached himself to Zup's leg. The hotshot, who had cropped red hair, freckles, and ZUPPIGER tattooed across his stomach in ornate script, pulled off his Seattle Mariners cap and twisted it onto the two-year-old's head.

With Jacob still fastened to his leg, Zup hugged Janae, who was grilling chocolate chip pancakes and bacon at the stove, and poured himself coffee. He took a seat at the table with Bunch.

"Boys are good, I see," Zup said. "And that one?" He pointed to Janae's stomach. Janae had recently learned that she was having another boy.

As the adults caught up about parenthood, Ben, the youngest, perched on the counter in the kitchen and Jacob sat cross-legged on a pillow eight inches from the TV screen. He worked the better part of his fist into his mouth. The cartoons still blared.

"Jakey, dude, can you sit on the couch, please?" Bunch asked. Jacob was Janae's child from an earlier relationship, but before they married, two years earlier, Bunch had insisted on adopting him as his own son. Jacob looked at his dad for a long moment, then pulled himself far enough from the TV to spare his retinas. The boy had showed signs of autism, and the Bunches were still figuring out how to cope with the possibility that their son might have the condition.

"It's hard enough having two normal kids," Bunch said. "And now we've got another coming . . ."

Zup was single but empathetic. At twenty-seven, he couldn't yet imagine raising kids or the turmoil the Bunches must have felt. He knew what it was like to get hard news, though. Not long before starting his rookie season in 2012, Zup's girlfriend at the time had taken her own life in the Prescott apartment they shared. He rarely spoke about that dark period of his life, and that morning at the breakfast table he steered the conversation to lighter topics.

Garret told the Bunches about his new girlfriend, and a story from a few nights earlier, when Garret had gone to a bar with Donut, Chris MacKenzie, and the young rookie Kevin Woyjeck. After a few drinks, Chris had started frisbeeing the rookie's hat across the bar. Woyjeck would pick it up and sit back down, only to have Chris rip the hat back off and once again fling it across the bar. Zup, who had a business degree from the University of Arizona, was a gifted writer and a quirky storyteller, even keeping a lighthearted blog, *I'd Rather Be Flying!* His recounting of the bar scene made Bunch feel like he was on the barstool, laughing with the other veterans as Woyjeck's hat floated across the pub.

Bunch had been with Granite Mountain since 2010, the crew's second year as hotshots, and he remembered the men back then being "nasty dudes who never took their yellows off. We were proud to be hard." Marsh's standards were high, and Bunch did everything he could to meet them. He was a quiet person to begin with, and nervousness scared him nearly silent for the first few months of his fire career. In an attempt to prove himself, Bunch once worked himself to heat exhaustion and had to spend an hour sitting in the shade. He returned to work that same afternoon.

His tenacity seemed to leave an impression on Marsh. Later in his rookie year, Bunch was arrested for drinking and driving, an offense severe enough to merit his firing. The only reason he kept his job was because Marsh convinced the department chiefs to give him a second chance. Bunch was nineteen at the time, and Marsh forgave his recklessness.

"It's nice to see you've got a little life in you," Marsh told Bunch.

The fact that Eric had stood up for him made Bunch even prouder to work for Granite Mountain. Marsh, though, remained an enigma to him. Bunch never felt he had a good read on his superintendent, but then, few people who worked with him did. One former colleague called him an "onion with layers he doesn't let most people see." He was typically serious, but he had a sharp wit and could be funny. The hotshots called Marsh's lingo "Eric-isms," and they included quips like "It's hotter than two rabbits screwing in a sock."

Marsh was a polymath and a bit of a snob. He took pride in understanding his pursuits well, and whether it was music, bikes, his tattoos, or his books and coffee (always locally roasted and shade-grown), Marsh enjoyed the finest things he could reasonably afford. His discerning taste gave Bunch the impression Marsh was cerebral but often lacking in the real-world experience to back up his ample theoretical knowledge. Whether it was mountain biking, riding horses, or firefighting, he knew all the right terms and the way to present himself. He also seemed to be searching for identity and affirmation.

The more seasons Bunch worked for Granite Mountain, the more he felt that under Marsh's command, the hotshots were always having to prove themselves. Some hotshot crews required their men to shave regularly, wear clean shirts, and keep their hair short. Granite Mountain was this way, and to Bunch, who was in the minority of the guys on the crew with no aspirations of becoming a structural firefighter, the crew's straitlaced vibe seemed at odds with the realities of a job that required digging line in hundred-plus-degree temperatures, camping out for two weeks at a time, and doing so with scant access to running water. By the end of the 2012 season, Bunch, who'd been raised on a ranch and spent his childhood riding bulls, had grown tired of "dressing up like a schoolboy to work in the woods."

That winter, he applied to two federal hotshot crews, in hopes of finding a crew culture that aligned more closely with his love of the woods. Then Janae got pregnant again. She didn't want him to go back to hotshotting at all. Raising two boys while he was away on fires was already testing his family, but planning for a third child changed everything.

Money was tight, and the Bunches had few options. With a high school education and no job experience outside of the hotshots, Bunch knew that fighting fires was the best way he had to make money quickly. If the crew hit a good assignment, he could save $2,500 in two weeks. Working for Granite Mountain would keep him close to home when he wasn't on fires. He and Janae decided that the best option was for him to return to the crew for another season. He'd leave just before Janae's due date. Marsh agreed to let one of his best sawyers leave early, and they set Bunch's last day for the first week of July.

PART TWO

THE
DEVOURING
FIRE

CHAPTER 6

THE DYNAMITE AND THE WICK

America's fire season usually starts in the Southwest, where desert winds and warming temperatures have primed the brush and pine forests for flame by early May. Wildfires then follow summer's heat counterclockwise, moving north and west up through Colorado, Utah, and Nevada before establishing themselves in the big timber in Idaho, Montana, and the Pacific states by August. Come October, the season has begun to slowly wind down in much of the West, but only now does it get going in Appalachia's oaks and the great spruce forests of Minnesota. Some years, an early storm drowns the sparks by November. Other years, the season ends with a flash in December, when the Santa Ana winds rake across Southern California's San Bernardino Mountains and push fires into Los Angeles and San Diego suburbs. But in the driest seasons, wildfires can burn straight through the winter.

The progression of the fire season is tracked out of Boise, Idaho, at the National Interagency Fire Center. Effectively, NIFC is the Pentagon of wildland firefighting. It dispatches nationally shared resources to the forty-five thousand to ninety thousand fires that ignite in America each year. NIFC, or simply "Boise," as it's often called by firefighters, sits adjacent to the airport and has the look and operational feel of a military base. The fifty-five-acre compound of drab buildings houses staff from the eight federal agencies—the Forest Service, the Park Service, the Bureau of Indian Affairs, the Bureau of Land Management, and others—that collectively manage more than seven hundred million acres.

One building on campus is dedicated to managing one of the world's largest civilian caches of radios. There's enough communications equipment to support thirty-two thousand firefighters. On-site inspectors ensure that the agency's fire vehicles are up to snuff near a smoke-jumping loft where soon-to-be-airborne firefighters sew and repair parachutes. There's a Costco-size warehouse that stores an enormous volume of fire equipment, and on the campus's perimeter waits a Boeing 737 that the fire agencies contract to fly firefighters to the most pressing blazes. There are no flight attendants on board; passengers must provide their own midflight snacks.

The brain of the wildland firefighting organization resides in NIFC's command center, a nondescript glass-paneled building in the center of the compound. Every morning during peak fire season, the fire directors of each agency meet around a long table surrounded by pictures of air tankers and flames. The directors' task: set the day's National Preparedness Level, a sliding scale that takes into account the percentage of available firefighters committed to blazes and the likelihood that new fires will spark. On one end of the spectrum is Level I—ample resources, few fires—and on the other is Level V: many fires, few available resources to fight them. In 2002, the directors placed the national preparedness at Level V for more than sixty days. That year, one battalion of troops from the Department of Defense was trained to fight fires, and firefighters from Canada and Australia were enlisted to add more able bodies to the effort. Over twelve months, more than seven million acres in all fifty states burned.

To compartmentalize how fires are managed nationally, NIFC, which was founded in 1965 in response to a need for cohesion among the many agencies battling wildfires, broke the country into eleven different fire regions. The boundaries were drawn to meet the particular challenges a certain area's fire regime poses to firefighters. Containing blazes in the vast wilderness of Alaska (Region 10) requires a response different from what's needed to fight fires in the brush, high winds, and dense subdivisions of California (Region 5). Geographic Area Coordination Centers were placed inside each region to act as the area equivalent of NIFC. At each GACC are smaller, though still incredibly extensive, caches of firefighting equipment. Each GACC staffs analysts and scientists who study the weather and fuel conditions to predict the region's fire severity for the day, week, month, and year.

Boise compiles all eleven regions' fire severities into a national forecast that's used to place resources in the regions that, because of drought or predicted thunderstorms, need firefighters most. NIFC and the GACCs make up the arm that moves the hammer in a high-stakes game of whack-a-mole.

Through the winter of 2012–13, the Southwest was dry enough to kindle a fire at any point, but the sparks didn't arrive until May 30, when a downed power line ignited a blaze near three small fishponds in Pecos, New Mexico. Twenty-four hours later, two thousand acres of ponderosa forest were burning and the season had begun.

Chuck Maxwell, a fire weather meteorologist, watched the blaze spread from his office at Albuquerque's Southwest Coordination Center, the home base for Region 3. As the lead fire weatherman for Arizona, New Mexico, West Texas, and the panhandle of Oklahoma, Maxwell indirectly determines where to send resources and when to order more for an area that covers 250,000 square miles. He's good at his job. Over the past fourteen years, he's predicted with almost 70 percent accuracy where and when to expect the biggest of the region's five thousand annual fires.

May 31 was the first time in 2013 that the SWCC felt hectic. To find the resources needed to contain the fire developing outside Pecos, a couple dozen interagency dispatchers called BLM, Forest Service, and Park Service offices throughout the region. Whether the dispatcher's specific job was logistics (caterers, Porta-Potties), communications (radios), operations (engines, hotshot crews), or aviation (air tankers, helicopters), the question they ask at a local, regional, or national level is almost always the same: What can your fire district spare?

Once the SWCC had exhausted the supply of available firefighters in the region, the resource order was bumped up to NIFC, which passed the request to the ten other geographic regions. Montana (Region 1) and the Northwest (Region 6) were already sending crews. By midmorning on the 31st, some two hundred firefighters and support staff were already en route to the new fire. Still, Maxwell thought they'd need more. A Red Flag Warning had been issued for that day, and dry and windy conditions were likely to fan the flames.

In his usual uniform of denim on denim, Maxwell stood before a fifty-five-inch monitor, watching a radar image of digital clouds drift over a black-and-white terrain map of the West. From this same spot, he'd watched more than a dozen blazes set and reset the record for the largest and most destructive fires in New Mexico and Arizona history. One fire he tracked developed into the largest recorded in the history of the lower forty-eight states.

That morning, dozens of wind barbs—tiny witch's-broom-shaped icons that indicate wind speed and direction—textured Maxwell's monitor in the area of the southern Rockies. Warm air on the desert floor colliding with cool air in the mountains formed an atmospheric instability, and already wind, nature's tool for creating equilibrium, was gusting at seventy miles per hour above the fire site.

Maxwell, who had spent two decades studying Southwest weather, wasn't too concerned about the blaze burning outside Pecos. It was the fires that hadn't started yet that really worried him. To predict where and when big fires are most likely to spark, Maxwell factors in drought, temperature, precipitation, the amount of fine fuels (like grass and brush), and the monsoon, a pulse of moisture that settles over the Southwest from early July to September and often ends the Southwest's fire season. Accounting for all of these variables, Maxwell saw enormous potential for large fires in 2013.

That summer's drought was the determining factor. The entire

Southwest was in year thirteen of a hot-and-dry spell that scientists were already comparing to the 1500s, when drought killed most of the region's forests. It's for this reason that few trees older than five hundred years can be found anywhere in the Southwest. The current dry spell was exhibiting the potential to be just as severe. Climatologists at Los Alamos National Laboratory predict that by 2050 we'll see a three-to-five-degree temperature increase over the Southwest and an even more parched clime. The effect a warmer and drier environment will have on fire seasons is predicted to be dire. Some scientists project 142 percent more acres burned every year in the Southwest and 50 percent more nationwide.

Because of the drought, Maxwell forecast that fires could become big and unruly almost anywhere in Arizona and New Mexico. The four other factors helped refine his predictions. The Prescott area concerned him greatly. The fall 2012 monsoon season had been wetter than average over the city, providing the area's wild grasses with plenty of water. By spring, the hills were cloaked in a vibrant green. As the summer warmed and the grass dried to straw, the fine fuel would become like a wick on dynamite, allowing flames to spread quickly between pockets of denser fuel like trees and brush. The combination of drought-cured forest and copious light and flashy grasses created potential for the most explosive fires in Arizona's history. But as Granite Mountain knew from the prairie burn just a few weeks earlier, the fuels weren't dry enough to support volatile blazes yet. Maxwell knew that the Prescott area wouldn't come into its prime until late June.

In the meantime, fire season in New Mexico's forests had already begun. The 2012 monsoon that watered the grass near Prescott missed much of New Mexico. So did the snow. The parched winter led to very little grass growth—no wick—but it turned New Mexico's forests into a tinderbox.

"Got a smoke report on Forest Road 105. Would like 301-UT to respond. Also 302 and Chase," radioed Todd Lerke to the local dispatch center as he raced his pickup toward the start. Lerke was the Jemez Ranger District's assistant fire management officer for New Mexico's Santa Fe National Forest.

Shortly after noon on May 31, high winds knocked down a power line in a subdivision six miles north of Lerke's office in Jemez Springs, a predominantly Native American and entirely adobe town at the base of the eleven-thousand-foot Jemez Mountains. An hour out of Albuquerque, the state had a new large fire. Called the Thompson Ridge Fire, it was northern New Mexico's second start in as many days.

Lerke ordered every resource in his New York City–size district to respond. Considering the area they covered, Lerke's force was tiny: ten firefighters and two small engines that together carried five hundred gallons of water—a little more than the average U.S. household uses each day and a fraction of what's used to handle most fires. It was all Lerke had. After getting the day's Red Flag Warning, he put his engine captains on high alert. But it wasn't the weatherman's prediction that put Lerke on edge as much as the forest itself. Already, pine needles crackled underfoot. He'd never seen the trees so dry so early in the season.

By the time Lerke arrived on scene, the Thompson Ridge Fire had jumped from the ground to the crowns of the tallest ponderosa pines. Flames reached above the forest. A woodpile beside one home in the dispersed community had caught fire. So had a railroad tie beside another. The smoke billowing from the forest was already dark like burning coal.

It's Las Conchas all over again, Lerke thought when he first saw the column. In 2011, the Las Conchas Fire had burned 156,000 acres of forest, and for a short time it was the largest blaze in the state's history. As massive as it was, the fire's intensity was far more disconcerting. For fourteen hours straight, Las Conchas torched an acre of pines every 1.17 seconds. That's sixty city blocks of forest up in flames in less time than it takes to microwave a Pizza Pocket. Las Conchas burned for five weeks. By the time it was over, the government had spent $48.3 million fighting it, sixty-three homes had burned, and a scar that looked radioactive—miles upon miles of devastated forest—spanned nearly a third of Lerke's fire district. The thought of another Las Conchas terrified him. He ordered additional helicopters, air tankers, more engines, and two bulldozers to Thompson Ridge.

As incident commander, Lerke separated himself from the action and formulated a plan for keeping the fire from marching straight into the nuclear research center at Los Alamos, a town of eighteen thousand people that sits just ten miles from where the blaze had ignited. Best case: His initial attack forces would catch Thompson Ridge in the next few hours. Barring that, he intended to form a box around the fire by linking together a series of logging roads in the area. Lerke parked in a meadow in sight of the fire and spread out a map on the hood of his truck.

A few seconds later, a propane tank exploded. When flames heated the 250-gallon tank of one of the nearby houses, a pressure-release valve opened to prevent the tank from detonating like a pipe bomb, but the sudden release of flammable gas near open fire resulted in a thirty-foot flame that looked and sounded like the roar of a fighter jet's afterburner. It vented every few minutes.

By then, the Thompson Ridge Fire had been burning for less than an hour and spanned nearly twenty acres. Flames had crossed from the pines near the houses into a field of oak shrubs on the flank of eleven-thousand-foot Redondo Peak. Lerke recognized the field as his last best shot at stopping Thompson Ridge before it became truly disastrous. Oaks hold more moisture in their leaves than pines do in their needles, and in New Mexico the shrubs rarely burn until late summer. To reinforce the natural barrier, Lerke requested that an air tanker from Albuquerque drop a strip of retardant on the top of the brush field. The drumming of propellers signaled the massive prop plane's approach from the south. It dipped to the height of the trees—so low that the crowns bent in the rotor wash—and dropped the retardant directly perpendicular to the fire's spread. If the blaze was going to escape, it had to burn through a field of oaks coated in three thousand gallons of retardant first.

From the meadow, Lerke watched through a veil of gray smoke as bursts of orange swept uphill toward the brush field. Bush to bush the fire jumped, the flames growing from ten to twenty to forty feet as they climbed upslope. It was not what Lerke wanted to see. The dampness of the oaks clearly hadn't slowed the burn. Nor had the retardant. The fire ripped through the line without so much as a pause. *Las Conchas all over again,* Lerke thought once more.

That afternoon, the men of Granite Mountain were still at their station in Prescott, but the season's slow start was about to come to a harsh and rapid end. They'd be fighting fires nearly without pause for the rest of June.

CHAPTER 7

TO THE EAST

"Where are you going this time?" Janae asked.

"New Mexico," Bunch said. The crew was already on the road.

Janae wasn't pleased. One week, two weeks, maybe even three, alone with the boys, plus the pregnancy—they needed the money, but she refused to feign enthusiasm about her husband's departure. At least Bunch wasn't going to a fire. Yet. For now, the SWCC had ordered Granite Mountain, along with a handful of other regional crews, to Albuquerque so there would be resources close should any of New Mexico's quickly growing fires need additional firefighters.

Staging, as assignments like this were called, wasn't an ideal job. It meant traveling to a distant station and sitting and waiting for a fire to start. There was no chainsaw work to pass the time, no freedom to drink beer after work, and no family nearby. In other words, staging was boring. Still, the hotshots didn't mind. They were finally leaving Prescott. In the Alpha buggy, the hotshots bent over their phones and sent flurries of text messages to friends, girlfriends, wives, parents— *We're going! First assignment! Drinks on me when we get back!*

East of town, the Prescott Valley resembles Napa without the vineyards. Its rolling grasslands are studded with agave and oak trees that are short, squat, and far older than they look. Scott, who sat in front of Grant, was pleased to finally be on the road, even if this wasn't the best time for an assignment. His older sister, Joanne, was pregnant and due any day, and he'd just adopted a new puppy, Riggs. Then there was Heather. He thumbed out an anxious text.

"Well baby, we finally got the call. We're heading to New Mexico right now. I'll miss you so badly. I'll call you tomorrow if you have time to talk later, we'll be driving for a while. I love you so much, sexy baby."

Heather Kennedy, a cop for the City of Prescott, was heading to a call on downtown's Cortez Street when she got his text. Scott slid low in his seat and they talked for half an hour. In the background, Heather could hear the drone of the buggy's heavy tires on pavement and the guys giggling at some probably lewd joke. She and Scott talked in the cooing tones of young love about things romantic and not: how maybe they should have waited to get Riggs until after the season ended; who was going to feed the dogs; how much they already missed each other.

Heather was Scott's first long-term girlfriend. She was slight, with cropped blond hair and a no-nonsense manner that indicated a toughness her stature didn't. They'd met a year and a half earlier on Match .com. She liked Scott's profile picture. He'd shot it himself while sporting a two-week traveler's beard in Thailand, Laos, or maybe it was Cambodia—somewhere tropical, on one of his many winter wanders. A monkey sat perched on his shoulder, and Scott wore a skeptical smile. He looked worldly.

Their first date was in a coffee shop, and Scott arrived late because one of Heather's colleagues had pulled him over for speeding, a bad habit of his. When he walked in, Heather was waiting for him in a booth toward the back. She faced the door as she'd been trained, so as not to be vulnerable to whatever threat walked in, and she recognized Scott immediately from his picture. She waited to say hello. She wanted to watch how he moved. Scott took off his sunglasses and craned his long neck around the coffee shop. He was nice to look at:

a head taller than her, thick red hair swept to the left, the lithe build of an endurance athlete.

It was formal and first-date awkward, until Scott mentioned getting pulled over. He remembered the officer's name, a sharp observation for somebody not in law enforcement, but she didn't let on that she was a cop. Guys always acted weird—threatened, even—when they found out. Equally coy, Scott said nothing about being a hotshot. He didn't want a "fire bunny," one of those girls who date only firefighters. They clicked over a shared interest in firearms.

Scott noticed the way Heather sat favoring one hip.

"Are you packing?" Scott asked her. She was wearing a Glock 19 on her waist. He couldn't believe it.

"No way!" he said, and lifted his shirt to reveal the same gun. At the time, he was working part-time at a gun shop and considered ownership of a wide range of guns a wise investment (a belief Heather and Scott's mother never shared). The Glock 19 is a popular sidearm, and both knew that carrying the same gun wasn't an exceptional coincidence, but they took it as a good sign.

Around the time a first date is supposed to end, Scott asked Heather if she wanted to go for a hike. They went to a lake by a formation of oddly shaped rock chimneys near town. She blew her nose onto the ground without thinking, then blushed. He pretended not to notice, then taught her how to hike with the long strides of a hotshot, bent slightly forward with her weight over her feet. He admonished her for swinging her head too much.

They dated on and off for the rest of the winter, and when fire season came around that spring, he returned to hotshotting. Scott's parents—a construction worker and an English teacher—had raised him to appreciate the outdoors. He hiked, rafted, snowboarded, and backpacked any time he had the chance. Working in the woods came naturally for Scott. For five fire seasons, he lived out of the back of his Toyota pickup while running chainsaw for the Payson Hotshots, up in the White Mountains to the east. He called hotshotting his dream job. The eight months of intense labor paid for his winter travels. He loved the hard work, the adventure, the brotherhood. But the first year he dated Heather, his job put stress on their untested relationship.

They'd been close that winter, finding and creating time to shoot guns, hike, and cook together, but Heather began to feel that Scott was more a friend than anything else. Raised Christian, he'd been a bit of a prude, too. Their relationship lacked romance, and that felt even more pronounced when he was away for weeks on fire assignments. After a few other guys showed interest, she decided to end it with Scott. When he finally called her during a fire assignment, she told him it was over. He'd actually laughed. Not because he thought it was funny. There just wasn't anything else to do. He was on an empty ridgeline, talking on a satellite phone, somewhere very far from Prescott.

They didn't speak again until he moved back home that winter. Heather called Scott first. She missed her friend and invited him to go to a shooting range in the hills outside of town. He brought his Glock 19. A squirrel popped out of the rocks, and he fired off three rounds but missed. He handed her the gun and she killed it instantly at thirty yards. He skinned it to make a chili dinner. That night, Heather asked if he was seeing anybody.

"Want to make sure you're not wasting your time, huh?" Scott said.

That February, he moved into her duplex apartment, on the corner of a busy street, with a yard big enough for her Belgian Malinois and a sign on the door with a picture of two German shepherds that said DOGS WILL BITE. Scott hung the squirrel's hide in the garage, where it served as a prompt for a funny story.

Despite his allegiance to the Payson Hotshots, Scott applied to Granite Mountain over the winter of 2012 to stay closer to Heather. The job had the added benefit of moving him closer to his goal of becoming a structural firefighter. Scott filled out what he called "the most complicated and in-depth hotshot app I've ever seen" during a break from cleaning his M4 assault rifle at the kitchen table. He was one of Marsh's top candidates.

Heading out for the first assignment of the season, Scott must have remembered getting dumped the last time he'd called Heather from the fire line. He recalled a conversation they'd had just weeks before the crew was sent east for staging.

"If I get a call, I have to go," Scott had told Heather before they adopted the puppy together. "Hotshotting is my job, babe. You know that."

"I get it, Scott," Heather had said. "I'm all in."

Not long after leaving Prescott, the buggies dropped down into the cottonwoods along the Verde River, a muddy stream that drains the twelve-thousand-foot San Francisco Peaks, to the northeast. They caravanned past Montezuma Castle, a forty-five-room dwelling the Sinagua people built into a sandstone cliff wall nine hundred years ago, and climbed into the junipers and piñons along an old backcountry highway that parallels the Mogollon Rim. On one side of the two-thousand-foot escarpment were the meadows and pine forest that cloak much of the Colorado Plateau; on the other, the Sonoran Desert. Near the cliff's edge, the road turned to dirt and Steed's truck skipped a bit transitioning from the blacktop to the gravel washboard. He didn't slow, and a cone of dust billowed out behind.

Granite Mountain was no longer going to Albuquerque. Barely out of Prescott, the SWCC had reassigned the crew to initial-attack a new fire burning just north of the Mogollon Rim in Arizona's Coconino National Forest, which sprawls over nearly two million acres of mountains, ponderosa-covered plateaus, and red-rock formations in the north-central part of the state. The new fire had started that afternoon during what law enforcement described as a spirit quest gone wrong: A camper had lit and lost control of a bonfire along a ridgeline overlooking a popular lake between Flagstaff and Phoenix. The escaped fire, called the Hart, was burning in a dangerous patch of dead trees, and fire managers in the Coconino National Forest wanted it out before it became a serious problem. Before nightfall, Scott and the rest of Granite Mountain would be cutting line.

From his seat, Scott looked out over the Mogollon Rim, out past the sandstone cliffs in the foreground, over the ponderosas and firs sweeping down through the rim's apron, and onto the desert and a layer of dust clouding the horizon in the distant south.

He knew this country well. It was where he'd fallen for hotshotting. A few years earlier, on an assignment not far from where Granite Mountain's buggies were bumping down the road, he'd shot a short video. In it, Scott's wearing his hard hat and yellow Nomex shirt. He points the camera at himself and says, "Sitting here in the shade watching two of my favorite things develop." He pivots the camera to flames in the foreground. "Fire." Then pans to a series of dark-bottomed clouds. "And weather."

CHAPTER 8

BAD MEMORIES

The prospect of fighting fire on the Mogollon Rim raises the hackles of any hotshot who has been around long enough. The pines grow thick, the escarpment is predictably windy, the terrain falls precipitously, and—because these features also add to the cliff band's astonishing beauty—thousands of vacation homes are scattered throughout the forest. But the Mogollon Rim's notoriety stems almost entirely from one particular day when a blaze, the Dude Fire, became a meteorological anomaly. It still haunts the profession today.

June 26, 1990, broke every heat record in Phoenix history. It was 122 degrees. The airport canceled flights because planes had never been tested in such extreme conditions and the tires might melt—nobody knew. A regionwide heat wave had hit during a drought like none seen in the Southwest since the 1950s, and forecasters predicted June 26 would be the most volatile day of Arizona's fire season. It exceeded expectations.

When plants burn, the water held in their tissues vaporizes, and meteorologists studying the blaze years later calculated that by around noon the Dude Fire had sent more than a million gallons of water skyward. Thirty thousand feet above the flames, the heat of the fire and moisture combined to form a supercharged thunderstorm that perched directly over the blaze. A wildfire creating its own weather is a relatively common phenomenon, but what happened around 2 P.M. on June 26, when the fire was more than a thousand acres and burning at its peak intensity, was exceedingly rare and dangerous: the thunderstorm and smoke column collapsed.

By noon on June 26, the Dude Fire had spread across three square miles of ponderosa forest, with one flank burning within a few hundred yards of the eighty homes in the Bonita Creek subdivision, forty miles north of Payson. As always, after personal safety, the firefighters' first priority was protecting homes. The incident commander recognized the blaze as far too intense to attack head-on and instead opted to herd the fire, using five hotshot crews and two Type 2 Initial Attack crews to attempt to contain the blaze.

Seen from above, the lines used to steer the Dude Fire around the Bonita Creek subdivision formed an incomplete *H*. The road atop the Mogollon Rim contained the blaze's left flank. Some two thousand feet below this line and running parallel to the rim, a firebreak the Civilian Conservation Corps had built in the 1930s, called the Control Road, contained the right flank. Walk Moore Canyon sat perpendicular to the two lines, between the houses and the flames, and connected the legs of the *H*.

Though Walk Moore Canyon was thick with ponderosa pines and cured logs, the incident commander planned to use the dry and rolling wash to push the fire around the subdivision. When the Dude approached the homes, firefighters would intentionally burn out the canyon, thereby blackening the fuel between the fire and the houses and forcing the flames into the unburned forest just beneath the rim. The key to the operation lay in an old jeep trail threading through the canyon.

The day before, a bulldozer had widened the trail into a fire line, and seven crews piled into the upper third of the mile-long wash. Around noon on the 26th, the hotshots closest to the top of Walk Moore started burning out while the other six crews widened the line and watched for spot fires. The operation didn't start well. Embers from the burnout immediately rode plumes of smoke across the line and ignited multiple spots just beneath houses in Bonita Creek. One spot took an entire hotshot crew to control.

They didn't know it at the time, but far more dangerous than these spot fires was the effect of the hotshots' burnout. It poured additional heat, moisture, and energy into the smoke column, which now billowed into the base of a fully developed thunderstorm. Firefighters coming to the Dude from six states could see the mushroom cloud, a pillar of gray boiling into the otherwise clear sky, from more than a hundred miles away. Inside Walk Moore Canyon, the day looked overcast.

"If it's burning like this now, imagine what it's going to be doing at one P.M.," one hotshot told his superintendent. Upon entering the canyon, another simply quoted Tennyson's poem "The Charge of the Light Brigade": "Into the valley of Death / Rode the six hundred."

The Perryville inmate crew was one of the seven crews tasked with the burning operation in Walk Moore Canyon. Five hotshot crews worked above them, and one, the Navajo Scouts, below. Shortly after 2 P.M., the Perryville inmates huddled tightly around a ten-gallon water can just delivered via ATV. They'd been building fire line nonstop since 1 A.M., and most hadn't had a drink in hours.

Unlike hotshot crews, Perryville used a rotating cast of superintendents, and on June 26, Dave LaTour, a professional firefighter from Tucson with a Tom Selleck mustache, commanded the crew. Beneath LaTour were seventeen inmates and two guards—Sandra Bachman and Larry Terra—who also acted as squad bosses.

In most cases, inmate crews aren't used for frontline assignments, but when the Dude broke, the incident commander made an exception and requested Perryville by name. He'd recently worked with the inmates on a fire outside Prescott and considered the crew capable and hardworking.

Many firefighters hold inmate crews in similar regard. After serving their sentences, it's not uncommon for former inmate firefighters to get jobs on wildland crews. Southern California first used prisoners on the line in 1949, and other states quickly adopted the program. Inmates proved an effective and inexpensive source of labor. In 1990, Perryville prison paid firefighters forty to fifty cents an hour to do largely the same job as hotshots. Working on the crew was more about pride than payment, and the prisoners competed fiercely for positions on Perryville.

"You were somethin' special. You know what I mean?" said Perryville firefighter Steven Pender. "It was like, *that's* the fire crew. You gotta be elite to get on that."

Predictably, Perryville's crew members were a varied lot. Among them were Curtis Springfield, an artist in for assaulting his girlfriend; Geoff Hatch, convicted of burglary and theft; and James Ellis, serving a twenty-year sentence for manslaughter. Each was immensely proud to be on the crew. To train for the season, the inmates would gather under the prison yard's razor wire and floodlights and spend hours doing push-ups and wind sprints. Recently, they'd been plotting a course to earn Perryville hotshot status, but their dream would never be realized.

Moments before the column collapsed, a frightening calm settled over Walk Moore. Perryville corrections officer Sandra Bachman was near the water jugs when she felt a light rain. The Dude was her first fire. She held her palms out and looked up into the ashy spritzes. Raindrop-size craters appeared in the dust around her. "I hope we get more of this," she said to an inmate.

The other crews in the canyon noticed similarly odd things. One superintendent said, "The fire line and the world became deathly quiet." Another watched tendrils of smoke that had been bending uphill with the daily winds suddenly stand straight up. The change in conditions made them uneasy, and all five hotshot superintendents pulled their crews back into a bulldozed safety zone—an area cleared of flammable materials. Only the Navajo Scouts and Perryville remained in the canyon.

A helicopter pilot, Dean Battersby, felt the column collapse first. Battersby was delivering brown-bag lunches to the crews working along the top of the Mogollon Rim. The blades of his JetRanger *thwapped* beside the column, and inside the smoke, he could see burning bark and branches thousands of feet above the fire and floating upward in the rising heat. The distinction between the smoke and the clean air was crisp, like the seam that forms at the confluence of a milky glacial stream and a clear river.

The Dude Fire's column had reached thirty-five thousand feet into the stratosphere, only half the height of the tallest smoke columns ever measured, but significant nonetheless. At that elevation, the temperature approached minus sixty degrees Fahrenheit, and the moisture in the smoke froze into a halo-shaped ice cap that sat atop the enormous pyrocumulonimbus cloud as if it wore a yarmulke. Just below the ice cap, the water molecules released from the fire condensed into droplets. The thunderstorm had reached maturity. It started to rain.

Around the same time, the ice cap became too heavy for the cloud's structure to support it, and from the top down, the thunderstorm collapsed into itself. Somewhere in the center of the cloud's massive cauliflower blooms, the smoke rushing skyward met the rain and cold air sinking off the ice cap, and for a few minutes the wind at ground level stopped as the fronts battled. Then the cold air and precipitation overwhelmed the heat of the column and the energy from both was refocused down toward the crews fighting the fire below.

Battersby noticed the collapsing column first, at 8,000 feet. His helicopter jolted to a stop. The pilot let out a grunt as he was suddenly lifted off his seat. The hands of the altimeter pinwheeled—7,500, 7,000, 6,500 feet—and in his rearview mirror Battersby saw the lunches, carried in a sling load that usually trailed a hundred feet behind the chopper, suddenly float directly beneath the JetRanger. He was plummeting in a descending column of air.

As the ground rushed up at him, Battersby raised the helicopter's nose while simultaneously redlining the power. He instinctively banked the JetRanger into a U-turn—the direction he'd come from provided the fastest escape from the downdraft. As the helicopter came around, Battersby scanned the forest below for a soft spot to crash: *Trees, always look for trees.* Open ground and water have no give, but branches break.

The JetRanger jumped and shuddered as Battersby flew into the safety of the calm air just beyond the downdraft. He dried his palms, climbed up to the top of the Mogollon Rim, and delivered the brown bag lunches. By the time he finished, forty-mile-per-hour winds were stoking the Dude Fire.

Just a few miles away and not far from the lunch spot, LaTour was scouting for spot fires when the smoke surged into the canyon. The descending column hit the Dude Fire uphill of LaTour. The winds fanned every flank and tossed embers in all directions, but in Walk Moore the walls channeled and accelerated the gusts from forty to sixty miles per hour. The gray overcast light suddenly turned an eerie and muted orange as embers the size of pinecones rained down upon LaTour and kindled in the pine needles with a sizzling hiss. He raced back to find Bachman, Springfield, Ellis, and the rest of the crew.

They were still at the water cans when the fire exploded. It was as if a heavy stone had been dropped into an already overflowing pot. The fire rolled over the top of the forest and jumped the two-foot flames that had been backing down the canyon—the hotshots' burn-out. In an instant, entire stands of trees combusted, and embers were thrown far down the canyon.

"Get the fuck out!" somebody on Perryville yelled.

The crew were fleeing in a tight group before LaTour reached them. As he passed the forty-pound water cans, LaTour grabbed one—why, he'll never know. He put himself last in a long line of forty Perryville crewmen and Navajo Scouts now hurrying downhill toward the vehicles parked on the Control Road. Many ran. LaTour chose not to. He didn't want to incite panic. He felt it, though. From behind him, he could hear the roar of flames—half again as tall as the pines—and the trees, cooked from the inside, exploding like artillery fire.

Meanwhile, a squad boss and an inmate who had gone downcanyon to fetch water were still picking their way back toward the crew when they heard a wind they described as sounding like a locomotive. Three Perryville crew members sprinted past them. The inmates said nothing when they passed—nor did the next three. The squad boss asked one fleeing firefighter, "Are you the last person?"

The inmate didn't answer. Instead, he started yanking at his fire shelter while running. In his fluster, he stubbed his toe on a rock and flew headfirst into the creek bed, bruising his ribs. Five Perryville crew members hustled past him as the flames and the panic closed in. The squad boss ran. Bachman fell down. James Denney, an inmate in for burglary, picked her up. An elk with fire just feet from its rear hooves hammered past them down the canyon. One crew member didn't put down his Pulaski. Another ran with a chainsaw. Burning debris landed under one man's shirt, and without time to pull out the embers, his chest burned as he sprinted beneath a ceiling of flame.

Greg Hoke ran in the middle of the pack—tenth in the line of twenty Perryville firefighters. Just as those ahead of him rounded a bend, a tongue of fire swept across the dozer line. Hoke, aghast, stumbled to a stop feet away from the flames blocking his escape. He turned around, ready to run uphill, and watched in horror as a second finger of flame crossed the canyon between him and LaTour, Bachman, and the eight other members of Perryville attempting to escape. Hoke presumed they were dead. He stood alone in the middle of a dry creek bed with flames curling around him. Out of time and options, he threw his pack aside and deployed his fire shelter.

Almost a mile downcanyon, an engine crew was eating lunch where the Control Road crossed Walk Moore. To support the burnout operation, they'd spent the morning running a hose-lay up the canyon and along the fire line, but after finishing, the crew took it easy and watched the burnout. The captain described it as looking nice—controlled, slow, exactly the way the incident commander intended it to be.

Things started unraveling when the same eerie hush the hotshot superintendents had felt settled over the Control Road. Moments later, a fifty-mile-per-hour blast of wind slammed into the enginemen, and two-hundred-foot flames appeared on the western edge of Walk Moore. *Get in the trucks!* the captain yelled. The fire would be upon them in minutes.

With no time to de-rig the hose-lay, the captain pulled out a knife and was madly sawing at the hose when the first wave of Navajo Scouts and Perryville inmates sprinted from the canyon and pulled themselves onto the sides of his engine. Around thirty firefighters clung to the vehicles, and every one of them seemed to be yelling, their voices quavering in panic—*Keep going! The flames are on you!* The engines started to pull out. They'd run out of time.

The last of the Perryville crew members to escape Walk Moore ran onto the Control Road through a wall of smoke; the flames were so close that they left second-degree burns on his neck. The inmate got himself onto the engine as it pulled away, and flames rushed across the Control Road. Had he been just thirty seconds slower, he too would have been trapped in the horror unfolding above.

Despite what Hoke thought, LaTour and the eight others hadn't been killed. As the inmate deployed his fire shelter, the firefighters uphill from him stopped before the flames hit and started retracing their steps. LaTour heard yelling, then saw his crew coming toward him. A cyclone of flame—roaring and billowing black—spun clockwise behind the men. He dropped the water can. *Get your shelters!* LaTour yelled.

Bachman's was snagged on something. One inmate stopped to help her as LaTour and the others continued their flight uphill. Some three hundred yards above the site where Hoke deployed, LaTour stopped. Flames blocked their escape route. The main blaze had nearly caught up to the spot fires downcanyon. Perryville had nowhere to go. LaTour gave the order to deploy.

Ten firefighters climbed under their shelters. At first, as they had been trained to, they remained optimistic. *We're Perryville. We're tough. We're going to make this!* they yelled.

LaTour, who was the last under his shelter, took one last look around: It wasn't the ideal place to deploy. Brush and trees grew tightly around the site, making it a dangerous place to try and weather the firestorm, but deployments are never planned. It was the best LaTour could do. He radioed out. "Perryville has deployed," he said, unaware of Hoke's position down the valley. Then, one by one, LaTour counted off the number of firefighters with him: "One, two, three, four, five, six, seven, eight, nine."

The screaming started when the flames hit. By then the column's collapse was complete. The winds exceeded seventy miles per hour. One gust momentarily lifted LaTour's shelter, and smoldering debris shot under the gap. His legs burned. Through pinholes in the aluminum he saw flames outside. He heard the horrifying crinkle of aluminum foil as the firefighters around him shifted. "Stay in your shelters!" LaTour yelled. "Stay on the ground!"

The winds raked embers across the canyon floor, and dark smoke filled the canyon. The shelters' aluminum exteriors became superheated. When one inmate touched the wall, it burned his elbow, even through his fire-resistant clothes. Lying beneath the shelters was torture, more than some could take.

Geoff Hatch, who'd deployed just below LaTour and the others, couldn't stand the pain any longer. He stood into the heat and stumbled uphill through smoke too thick to see beyond his feet. The embers bounced off his legs, face, boots, hands. Out of earshot from the crew, as pines burned around him, Hatch fell to his knees and pleaded for God to end it all.

At the deployment site, Curtis Springfield, who was near LaTour, screamed, "I can't take it anymore!" He, too, got up. A few inmates echoed Springfield's howls of pain. He made it 150 feet before collapsing on his back and rolling downhill over glowing coals.

LaTour again yelled—pleaded—through the roar of the wind, "Stay in your shelters!" His crew was dying.

The pain overwhelmed James Ellis and Denney. Both stood up.

When Denney shucked off his shelter, the wind stripped it from his hand. Screaming and tottering, he followed the shelter downcanyon. According to research conducted after the incident, Denney's clothes had reached 824 degrees before another inmate, Joseph Chacon, stood up, tackled his friend, and forced him beneath his still-intact sheet of aluminum. They were found dead, Chacon lying atop Denney, both men severely burned.

When the second flaming front came through, LaTour's shelter delaminated from head to toe. The right side of the aluminum exterior folded over onto the left and beat against itself like a wind sock. LaTour covered his face in the crook of his arm and put his back toward the thin sheet of cloth separating him from the flames. Experts would later determine that the temperatures inside LaTour's shelter soared to six hundred degrees.

LaTour and the others lay in the dirt for forty-five minutes before the Dude Fire cooled significantly. Outside, as the winds died down, the demonic roar of a thousand trees burning at once slowly calmed to the pop and crackle of countless campfires. Just uphill of LaTour were the three inmates who had stayed under their shelters and survived the burnover. One called to the superintendent, asking if it was okay to get out. LaTour said to stay in. If it still felt as hot as it did inside the shelter, the outside could only be worse. Help was coming. It had to be.

But help didn't arrive. The fire had pushed into Bonita Creek, burning most of the subdivision's homes, and the hotshots in the safety zone were waiting for the fire's intensity to die down. Realizing Perryville would have to save themselves, LaTour eventually emerged from his shelter. The forest had become a wasteland of black ash and red heat. The burned bodies of his crew members lay scattered around him. Some of the fallen inmates were completely under the shelters, others half exposed. Bachman was dead. She lay faceup and twisted, with the homes Perryville had worked to save scorched above her.

LaTour told the three surviving firefighters, those still under their shelters, to get up. "But don't look," he said. "A tragedy has happened. We have to get out."

Ten minutes after Perryville was burned over, three hotshot superintendents tried to make their way out of the safety zone and into Walk Moore Canyon to help. One of them was Paul Gleason of Oregon's ZigZag Hotshots. He'd already fought six blazes in which firefighters had died, but in his twenty-six years on the line, he'd never experienced anything as horrific as the Dude Fire.

Gleason and the others made it just 150 yards before they saw Geoff Hatch, the Perryville inmate who'd pleaded for God to take him. God hadn't, at least not yet. For more than half a mile, Hatch had stumbled up the dozer line through pulses of fire. Burns covered 60 percent of his face, and folds of skin hung off his body. When the superintendents found Hatch, he mumbled, "Those damn fire shelters didn't work."

A few minutes later, EMTs and a stretcher arrived on scene, but the fire behavior was still extreme, and when Hatch saw flames leap the canyon and rush toward him and his rescuers, he kept repeating the same six words: "It is going to get me. It is going to get me. It is going to get me . . ." Gleason thought Hatch might be right. The fire came so close that at one point he briefly considered telling the others to run.

"I had a shelter, and there might have been room for the both of us," Gleason said later. "Whatever happened next would be between me and Hatch and God."

It never came to that. The EMTs loaded Hatch onto the stretcher and rushed past the still-burning skeletons of destroyed houses and into the safety zone. Inside the clearing, hotshots, some weeping, were already clearing a landing pad for a helicopter. Hatch was airlifted to the nearest hospital. He faced a long recovery, but he'd ultimately survive. Others weren't so lucky.

———

Hoke, the first inmate to deploy, lay beneath his shelter a few hundred yards downcanyon from LaTour's group. During his deployment he had heard, amid the gusts of howling wind and the roar of flames, what must have sounded like an aberration. James Ellis, one of his crewmates who had abandoned his shelter, was outside. Ellis had stumbled almost three hundred yards downcanyon to where Hoke had deployed alone in the creek bed.

"My shelter didn't work," Ellis said.

Hoke stayed beneath his shelter. It was still far too hot to leave. "Get the water from my pack," he told Ellis. He heard his crewmate shuffling around.

"It's burned up," Ellis said. Then, just as quickly as he'd come, he walked away, leaving Hoke alone.

The next time Hoke saw Ellis was when LaTour and the three other survivors, who had been under their shelters for more than forty-five minutes, walked down the canyon with their shelters wrapped around them like capes—using them as shields against the heat. Miraculously, Ellis was with them: burned from head to toe and barely alive.

In a loose group, the six survivors walked out the dozer line. LaTour, exhausted and fighting the urge to sit down, again took his place at the back of the group. For a half-mile, they tripped and stumbled through burning stumps and white ash toward the Control Road. Ellis didn't make it.

A few hundred yards from the end of the canyon, he muttered, "I'm dead." Then he lurched toward a log, sat down, and died. The others walked on.

Gleason, ZigZag's superintendent, continued alone into Walk Moore Canyon, following the route Hatch had taken out. He returned to the dozer line. Through residual heat, Gleason walked down the line until he saw a man's body through a window in the smoke.

"I rolled him over. I put my cheek right up to his nose and his mouth to see if there was any air," Gleason said. "A tear rolled off my eye and fell on his cheek. I saw it. I thought it was his tear, and for just a fleeting second I thought that this guy was alive. And then it settled on me. I was in a canyon with a bunch of dead people."

Gleason tried to radio his crew to tell them he was all right and to relay that they had fatalities, but he couldn't move. The aging superintendent sat on his knees over the fallen and wept—for five minutes or fifteen, he didn't know.

"I love fire. I like lightin' it. I like fightin' it," Gleason said. "And I'm there with this guy, but this was not what I had signed up for. I'm trying to tell myself that this part doesn't exist, that I had seen Hatch before and that was as bad as it could get." But it wasn't. Gleason spent the rest of the evening loading six blackened bodies into bags.

CHAPTER 9

LESSONS LEARNED

Scott Norris had been on summer break from elementary school back when the Dude Fire exploded in 1990, but he knew the story. When Scott worked for the Payson Hotshots, Superintendent Mike Schinstock took the crew to the fire site to see how "a bad day ends," as he'd tell his hotshots. They walked the route that Bachman and five others had followed to their deaths two decades earlier. In the wake of the tragedy, the residents of Bonita Creek subdivision planted six cottonwoods among the bleached ponderosa stumps that now fill Walk Moore Canyon, and in the shade of those trees are six stone crosses, each one marking the place where a firefighter died.

Standing before the crosses, it's hard not to imagine the canyon engulfed in flames. Schinstock asked his hotshots to do more. He passed around a picture taken the evening of June 26, 1990. It showed burned and twisted bodies, smoke still rising from the ash around them. *Think about what choices the Perryville crew made and why they made them,* Schinstock asked of his men. *And what, if anything, you might have done differently to change the outcome.*

Since then, Scott had studied tragedy fires. At home on his shelves, beside the books he devoured during the rare quiet moments of summer—Cervantes's *Don Quixote,* Kafka's *The Metamorphosis,* Vonnegut's *Slaughterhouse-Five,* and the *Reloading Guide to Handgun Accuracy*—were copies of hundred-page investigations into firefighter deaths. State or federal agencies fund these reports, which are compiled by teams of dozens of subject-matter experts ranging from psychologists to weathermen and fire-behavior experts. They use dozens of witness testimonials to piece together the events that led to the fatalities.

Though these reports are filed every time a firefighter dies on the line, Scott's interest lay specifically in burnovers, the grisly deaths that occur when flames overtake firefighters. He knew the details of the Dude Fire and most of the other historically significant mass tragedies: the Mann Gulch grass fire that overran thirteen firefighters, including twelve smoke jumpers, in Montana in 1949; the Loop Fire that exploded and trapped twelve hotshots in a narrow gulch in California in 1966; the blow-up that sent a wave of flame over fourteen elite firefighters in Colorado in 1994.

Men and women have died on the fire line every year since federal agencies started keeping track of deaths in the 1930s. Yet an expectation remains among the Forest Service and the other largest employers of wildland firefighters that it can be a zero-casualty war. The simplest way to stop firefighters from dying on the line would be to stop fighting wildfires and let them burn unchecked in America's wilds. That outcome seems highly unlikely, given—among other things—America's many billion-dollar investments in fighting fires and the many billions in private property that would be destroyed if we simply stopped. Instead, the reaction by fire agencies to deaths has almost always been the same: provide firefighters with additional rules, better equipment, and a more comprehensive education.

"We can't really make the job safer with more policy and rules," says Jim Cook, a former hotshot superintendent who went on to help found the Wildland Fire Leadership Development Program, an institution that strives to make firefighting safer by improving leadership skills and decision-making on the line. "There's an expectation that we can have a fire season where no firefighters die. But there's no evidence to suggest that we can actually pull that off. What we do after each major incident is go back and examine why we had such a bad outcome. Each time, we get incrementally better at fire-line safety."

The tradition of institutionalized learning after mass-casualty fires dates back to 1956. After the rapid succession of three mass-casualty fires left thirty-nine young men dead, the Forest Service commissioned a task force of experts to investigate why so many firefighters were dying and "materially reduce the chances of men being killed by burning." The task force's findings were distilled into ten Standard Firefighting Orders:

1. Keep informed on fire weather conditions and forecasts.
2. Know what your fire is doing at all times.
3. Base all actions on current and expected behavior of the fire.
4. Identify escape routes and safety zones and make them known.
5. Post lookouts when there is possible danger.
6. Be alert. Keep calm. Think clearly. Act decisively.
7. Maintain prompt communications with your forces, your supervisor, and adjoining forces.
8. Give clear instructions and ensure they are understood.
9. Maintain control of your forces at all times.
10. Fight fire aggressively, having provided for safety first.

Shortly after the investigators issued the ten Standard Firefighting Orders, they released the "13 Situations That Shout Watchout"—specific, cautionary environmental and human factors often present in tragedy fires. By the 1980s, after ten more mass-casualty fires, the name had been shortened to just "Watchout Situations" and the list expanded to eighteen. A few examples:

> Unburned fuel between you and the fire.
> Cannot see main fire, not in contact with
>   anybody who can.

Wind increases and/or changes direction.
Terrain and fuels make escape to safety zones
       difficult.

The ten Standard Firefighting Orders and the eighteen Watchout Situations are known by firefighters as the Ten and Eighteen, and these rules of engagement provide the basis for all fire-line decisions. Most firefighters carry a laminated list of the orders in their breast pocket. Scott did. He could also recite the directives from memory.

Superintendent Mike Schinstock was disappointed when Scott told him he was leaving the Payson Hotshots. He saw in Scott the trappings of a career fireman and had started grooming him for a leadership position shortly after Scott started with Payson. By the time Scott came to Granite Mountain, his fire-line qualifications rivaled those of the crew's squad bosses.

Though he often said, "That's not how we do things on Payson," he was thrilled to get the job, and he came to greatly respect Steed in the months he worked under him. Scott didn't talk much about his previous experience. But his quiet confidence and amiable personality made him a natural role model for Granite Mountain's rookies. Kevin Woyjeck, like many of the younger guys on the crew, looked up to Scott. Woyjeck reminded Scott of a younger cousin—eager to the point of annoyance, a little brother he couldn't help but like.

Woyjeck had peppered Scott with questions throughout the season, but the queries reached a dizzying barrage before Granite Mountain left for its first extended fire assignment: *Seven pairs of underwear— that's too few? What, really? Too many? Extra bootlaces, hooded sweatshirt, beanie—need those for sure . . . right?*

Dusk had fallen over the Mogollon Rim when Granite Mountain pulled into a gravel cul-de-sac across from the Hart Fire. Scott, Grant, Renan, Donut, and the others on Alpha squad poured from the back of their buggy and took stock of their task: a ridge awash in orange.

At that point, the Hart Fire had spread across a couple of acres of pine saplings and long-dead trees—both logs and still-standing snags. Occasionally, flames climbed into low branches and—*whoosh*—a shower of embers arced skyward as a single tree combusted. The little timber burn was far more impressive than the prairie fire weeks earlier, but the Hart was still just a preview of the fires to come. A few shifts and Granite Mountain would be heading home. But the best thing about the assignment was that Granite Mountain was one of only a few resources on the Hart, and working outside the purview of an incident management team brought out the hotshots' lighter side.

While working an even more remote fire, in Minnesota's Boundary Waters in 2011, Granite Mountain canoed into their piece of fire line and spent a week camping out alone on a beach. Scott wasn't on Granite Mountain for the Minnesota trip, but he'd been on plenty of similarly slow assignments.

On Payson, he'd seen guys jump fully clothed into Pumpkins, the inflatable orange five-hundred-gallon water tanks that are staged along fire lines. He'd taken part in the impromptu competitions that develop between crews working the same isolated piece of line. On one fire, an elected representative from each crew packed an entire can of Copenhagen into his lower lip, sprinted around a predetermined course, and did as many push-ups as possible. The first to vomit lost. Scott himself had tried and lost the 4:4:40, a challenge to drink four quarts of water in four minutes and hold it down for forty seconds. He loved the antics on slow fires. The playfulness was a perk, but Scott also enjoyed the work.

That night, the hotshots constructed line steadily but without any great urgency. Between the reservoir and the creek, water surrounded the Hart on three sides. To complete the box around the fire, all Granite Mountain needed to do was punch in a quarter-mile piece of line from the lake, up and over the ridge, and back down to the creek.

It was the type of quiet and cool evening that kept hotshots returning to the job for decades. As darkness came to the Mogollon Rim, the humidity climbed and the fire pulled back into the heaviest logs scattered about the forest floor. Some of the men turned off their headlamps and swung their tools by the fire's warm glow. Bats strafed the lake, and the muffled thud of the hotshots' tools striking dirt could be heard from the campsites across the water. Granite Mountain finished lining the fire that night, and at around 2 A.M. the men spread out their sleeping bags and fell asleep under the blue light of a half-moon.

The next morning, Steed and Clayton tapped Renan to join them on a burnout operation. Renan's heart jumped. "To be a rookie with Steed and Clay? I was like, *Oh shit, this is awesome!*" he'd tell Grant later.

The three hotshots hiked along a cobbled beach beside the reservoir. At that point, the Hart Fire smoldered in the cliff bands above them, even less of a threat than it had been the night before. Still, spot fires always remained a concern. To widen the fire's black edge and ensure the blaze was controlled, Steed wanted Renan to burn off a corner of willows and grass between the completed line and the lake.

Usually stoic, Renan could barely contain his enthusiasm. Burning out with Steed and Clay was surely a sign that they were considering him for Rookie of the Year. The rest of the crew faced the drudgery of mop-up, and Renan had drawn the most fun job on the line— legitimized arson.

"All right, dude," Steed said to Renan. They were ready to burn out the Hart Fire. "You're up. Don't breathe the smoke in. It'll stay with you all day."

From his pack, Renan grabbed a fusee, one of the dynamite-like sticks used as road flares. He twisted the top, breaking the paper between the chemical fuel in the tube and the igniter, and uncertainly flicked the two slate ends together. The stick hissed, and a foot-long orange flame and a cloud of sulfur erupted straight into Renan's face. He coughed and gagged. Steed boomed with laughter and showed him how to use the fusee to start burns in the undersides of bushes, where, sheltered from the rain, the pine needles and leaf litter were driest. By day's end, the low-intensity fire inside the box was nearly forty acres—twenty times larger than when the crew had first arrived the night before. That knowledge gave Renan no small amount of pride.

While Renan was clearly enjoying himself, the fire was testing Grant. Up on the ridgeline between the creek and the lake, he and the rest of Alpha squad stacked smoldering logs into piles to concentrate and burn down the heat. The bone piles, as the hotshots called them, were scattered around the forest. For Grant, the work amounted to a game of pick-up sticks. He paced the area closest to the line, seeking out burning logs that were small enough to move and, with his tool and gloved hands, dragging them into various burning piles.

Ash and soot smudged Grant's face, and streaks of black colored his yellow shirt. Renan hiked up from the lake, his burnout now complete.

"Best day of my life," Renan said, beaming. "Seriously, dude. That was fucking awesome. I just got paid to light the side of a mountain on fire."

"That's awesome, man. That's so cool," Grant said. But it was clear he was bored.

Once the bone piling was finished, Grant started mopping up. He dug a quick pit in the ash near the fire line they'd built the night before. With his rhino, he scooped up a smoldering twig, dropped it into the hole, and, to ensure a gust of wind couldn't rekindle the embers, ground the wood into the cooler dirt until no remnant heat remained.

"I've been standing here in the sun and mopping up this fucking hillside all day," Grant said. He hadn't slept well the night before. At home, he always put a funny movie on the TV—*Tommy Boy* was his favorite—and fell asleep to the noise. Last night, when Granite Mountain finally did go to bed, all he heard was snoring so loud it sounded painful and millions of chirping crickets or frogs or whatever they were.

Another smoke tendril twisted up from the ash at Grant's feet. He covered the smoldering twig with more cold dirt and stirred again. Down below, four or five of the other hotshots whooped as they dove into the creek. The line was in, and the work that was left wasn't pressing. In small groups, Steed gave the men a few minutes to relax and enjoy the place. The veterans had been whispering among themselves about how cool it was that Steed was letting them swim—it wasn't something Marsh had ever allowed. Donut took the time to tie a rock to a length of parachute cord and toss the makeshift lure toward the crawdads teeming in the shallows. When one crawdad latched a claw onto the stone, its natural defensive mechanism, Donut lifted the crustacean from the water, grabbed its body behind its flailing pincers, and pushed the animal into his breast pocket. It stayed there until he released it later that afternoon.

"Dude, you gotta relax," Renan told him. "You've gotta find something to look forward to. What do you like? Not movies, but something you can do out here. Something to take your mind off of this shit."

"Hand-rolled cigarettes. I like those," Grant said. He placed the back of his hand just above the ash to feel for heat—nothing—then buried his bare fingers into the pit he'd been working. Only the remnant warmth of the coals remained.

"Get some, then. At the next gas station stop, buy a pouch and smoke 'em down. In the meantime"—Renan tossed him a can of Copenhagen. "But don't get addicted. Leah wouldn't like that."

In the evenings after their shifts, the crew went for dinner to the nearby station of the Blue Ridge Hotshots, a Forest Service crew based in the Coconino National Forest. Blue Ridge normally would have dealt with the Hart Fire, given how close it was to their station, but they weren't even in Arizona. The SWCC had sent Blue Ridge to New Mexico after the Thompson Ridge Fire began. Since it started, the blaze had moved progressively closer to the eighteen-thousand-person town of Los Alamos. Blue Ridge's empty station had cell-phone coverage, running water, and a cache of Gatorade, which, as was common among hotshot crews, Granite Mountain helped themselves to.

One night while relaxing at the station, Chris MacKenzie, Alpha's lead firefighter, received a nightmarish voice mail. His mom had a brain tumor. She was going to the hospital for surgery.

"She said she was going to be fine," Chris told Donut, his room-mate, and Clayton Whitted, Alpha's squad boss. "It's a routine proce-dure. I should just stay here. I need the money." In some ways, staying with the crew was easier than seeing his mom incapacitated in a hos-pital bed.

But Clayton insisted Chris leave to see his mom. "You're going, Chris," Clayton said. "You're leaving tomorrow and you're going to be there for your mom."

Chris left the crew that next morning and, down another key crew member, Donut stepped into his lead firefighter position. Granite Mountain stayed to work the Hart Fire for another couple of shifts. By most of the hotshots' standards, the fire had been an easy one. But a few days later, after another long and slow shift on the Hart, Grant cracked.

Outside of Blue Ridge's station, Alpha squad sat in the back of the buggy, eating the steaks the Coconino National Forest had provided for dinner. Moths bounced off the cab lights, and in the chill of the spring night, some of the men wore beanies and hooded sweatshirts. Steed expected they'd be heading home to Prescott in the morning. The Hart Fire was lined, mopped up, and now mostly cold.

Grant sat quietly in the back of the buggy while the other men chattered around him. Hotshotting is at once intensely social and oddly isolating. The men get no time alone. Anthony Rose, apparently bored, started calling to Grant from his seat up front. Whatever it was—a reference to Leah, a lighthearted insult about Grant looking every bit as tired as he felt—Grant didn't like it. He stared at the back of Tony's head. Grant agreed to play by the hotshots' rules, but he wouldn't do it at the cost of his dignity.

"You're picking on me because I'm the only person you can pick on," Grant said. "You're nothing but a bully."

A hush fell over the buggy. Rookies didn't usually talk back, let alone with vehemence, and Grant's tone of voice brought to mind that he'd been a varsity wrestler in high school. Tony turned around, surprised. Rookies got picked on; that was hotshotting's social contract. On Granite Mountain, it took a year for the men to claim ownership of the title of hotshot. To build pride in the crew and give men a reason to return after their first season, Marsh had instituted a policy that rookies had to wear different T-shirts than the veterans. Until they finished their first year, new guys were treated as second-class. Grant found it frustrating. He'd willed himself through most of his life's challenges—prescription drugs, wrestling, Steed's workouts. But his willfulness couldn't force obstinate hotshots to treat him with the respect he felt he deserved.

"We've all passed our critical training," Grant continued, his voice rising. "Right now, I don't need your shit." Tony, who had recently found out his girlfriend was pregnant, ratcheted up the tension. He smirked dismissively at Grant, until Renan set his plate down, stepped across the aisle in the buggy, and gently pulled his friend outside.

CHAPTER 10

IN COUNTRY NOT SEEN
IN DAYLIGHT

I t's a six-hour drive from the Coconino National Forest to Jemez
Springs, New Mexico, where the Thompson Ridge Fire was burning. That morning, just before the hotshots left the Hart Fire for
home, the SWCC dispatched Granite Mountain to the rapidly expanding blaze. The distance provided the hotshots with one of the
job's great pleasures: the road trip. With no work to do and many
miles of wild landscapes to pass through, the men could spend the day
lounging and napping until their next shift. On fire assignments, the
buggies became home.

Alpha's was tricked out. Last year's squad had all chipped in to buy
a flat-screen TV that still hung above the half wall that separated the
eight hotshots from the squad boss and the lead firefighter in the cab.
All the way from the Hart Fire to the Jemez, Alpha squad played movies. Grant loved it. He called Leah and left a quiet voice mail.

"Just calling to let you know we got reassigned to a place called
Thompson Ridge, New Mexico, outside of Albuquerque. Hoping
that we'll be able to stay in contact with you. Anyways, that's where
I'll be. If not, love you and miss you."

Then he promptly kicked back in the air-conditioning and let
Chris Farley's *Black Sheep* soothe his frayed nerves.

Bravo cultivated a more rustic vibe in their buggy. The squad, led by Bob Caldwell and the enormous Afghan War veteran Travis Turbyfill, included Bunch, Zup, and Woyjeck. Not yet a week into the first fire assignment, the back of the Bravo rig had the sweet but mostly sickly smell of body odor, woodsmoke, and rot. Empty Gatorade and water bottles rolled around the floor. Somebody had mounted an elk's antlers—flecked with drops of red retardant—above the half wall. Surrounding this proud centerpiece were postcards collected from every state where Granite Mountain had fought fires: Minnesota, Alabama, Washington, Colorado, Montana, Idaho, California, New Mexico, Nevada. Copenhagen chewing tobacco lids were taped to the walls. So were knives and forks lifted from restaurants, a horseshoe hung with the points up so the luck didn't run out, and a picture of a frowning cat that read, THIS IS MY HAPPY FACE.

What both buggies shared were the small cardboard boxes that sat on the windowsill beside each hotshot's seat. In them, the men stowed their personal items: books, decks of cards, magazines, sunflower seeds, instant coffee. For hotshots on assignment, gas station convenience stores were the source of all good things.

Granite Mountain stopped at one in the Navajo country of northeast Arizona. Dust and tumbleweeds drifted across dirt roads so rough they saw traffic only from locals in rusting American-made pickups. It was a hundred degrees, and hotter by the pumps. For many convenience store owners in the West, seasonal firefighters mark the change of seasons as much as warming days do. Firefighters are good for business. Cold drinks, hot coffee, sunflower seeds, ice cream, chocolate, and logs of chewing tobacco, each one holding five cans, flew off the shelves by the armful.

"Dude, put those generic chips back. Get me the goods," Bunch said to Grant, who was carrying a load of snacks to the counter. The previous month, Grant had challenged Bunch to play him in a game of disc golf. Bunch refused. The harder Grant tried to gain acceptance, the more the veterans pushed him back. Grant chided Bunch until he finally agreed to a match, but Bunch set the terms: Loser buys the winner the chips of his choice at every gas station stop for the rest of the season. Bunch won by six strokes.

Bunch grabbed a bag of Tostitos and cheese dip and piled them onto Grant's stack, which included Zig-Zags and loose tobacco for rolling cigarettes—something to look forward to. Satisfied, Bunch stashed his chips and dip into his cardboard box, where they'd stay until the end of their assignment. He knew that after two weeks of MREs and disappointing lunches of Wonder Bread sandwiches packed with an inch and a half of deli ham, Red Delicious apples, and canned and always lukewarm juice, Bravo squad would be craving a proper snack.

The Thompson Ridge Fire had grown to seventy-four hundred acres in the five days since it started.

After the blaze tore through the oak field that Todd Lerke, the first incident commander, hoped would contain it, the fire jumped the roads he'd mapped out as contingency lines. Flames then spread northeast up eleven-thousand-foot Redondo Peak and threatened the Valles Caldera, a national preserve of open meadows and forests. A few historic cabins now sat directly in the line of the fire's spread. The bigger concern, though, lay just five miles beyond the Valles Caldera: Los Alamos and its Department of Energy–run labs. Since 2000, two separate fires that had started in the Santa Fe National Forest had razed hundreds of Los Alamos homes and burned sensitive lab property, some of it rumored to house a uranium dump from World War II. The town dreaded the prospect of yet another large burn.

So did the Santa Fe National Forest. On June 2, the SWCC put Incident Commander Bea Day and her team of Southwest-based logistics, planning, safety, operations, and finance experts in charge of the fire. Nine hotshot crews were already on scene by the time Granite Mountain pulled into the fire camp outside Jemez Springs to get their assignment. Unlike the low-key Hart Fire or the one-shift-and-done prairie fire, Thompson Ridge required a robust fire camp to support the more than five hundred firefighters on scene.

Thompson Ridge was considered a Type 2 blaze, the nation's

second-highest priority. The numeric system that's applied to the National Preparedness Level—the country's readiness to attack wild-fires—is flipped on its head to classify wildfires. Type 5 incidents, which might be a single burning tree or a campfire that requires very few resources to control, are the lowest-priority blazes. The highest are Type 1 incidents: complex, multifaceted fires that can cost many millions of dollars and may require thousands of firefighters and logistical support personnel to contain.

Classification is determined not by a blaze's size but by its potential and complexity—towns endangered, major freeway closed, nuclear waste facility threatened. Where only a few observers may be needed to track the progress of a hundred-thousand-acre burn in Alaska's wilderness, a four-acre fire near the suburbs of California's San Bernardino Mountains may be deemed a Type 1 incident at its outset. Charged with orchestrating the most complicated firefights are incident management teams like Day's. There are sixteen Type 1 teams—the most qualified—and thirty-six Type 2 teams nationwide. Management teams can vary in size from seven to seventy experts, most of whom hold full-time jobs with federal or state fire agencies. When a national forest requests a team to manage the fire, a GACC or NIFC orders the militias of logistical and tactical experts up from their agency day jobs.

Since the Incident Command System was first developed, after a rash of blazes plagued Southern California in the 1970s, NIFC's rapid logistical deployment system has become so refined that the Federal Emergency Management Agency turns to the Fire Service, along with the National Guard and Army Corps of Engineers, to assist with disaster relief. Hurricanes Katrina and Irene, the *Columbia* shuttle disaster, and 9/11 all utilized NIFC's Type 1 incident management teams.

Thompson Ridge was by far the largest blaze Granite Mountain had seen so far that season. Camp was set up on a helicopter landing pad atop an open ridge miles behind the flames. Tents and trailers sprawled across a few acres. On-site was a high-speed Internet hot spot, a cell-phone tower, and a portable weather-monitoring system brought by a meteorologist assigned to cover only the fire area. There were semis packed with hoses, extra shirts and pants, pumps, Pulaskis, and fuses; and semis packed with cases of Gatorade and brown-bag lunches. Portable showers sat by the mess tent, where hotshot crews from Arizona, New Mexico, and Montana stood in a long line for cafeteria-style food served by caterers from Albuquerque.

The whole place smelled of dust and diesel, with the occasional acrid waft of urine from the Porta-Potties. Gasoline-powered generators hummed endlessly, diesel trucks rattled around camp, and a fine dust that fire boots and tires had ground to the consistency of talcum powder coated everything. If Steed had his druthers, Granite Mountain would spend as little time as possible in fire camp. Immediately after arriving on scene, Steed and the squad bosses left the men at the buggies and headed into the mass of trailers and tents for that night's briefing. A makeshift stage had been set up in the middle of the camp, and an oversize printout map of Thompson Ridge was hung on a whiteboard. The fire's perimeter looked like a lobster claw wrapping around Redondo Peak.

Day's team had divided the 557 firefighters on scene into a day shift and a night shift. Gathered around the map were the overhead from fourteen different engines and hotshot crews that, like Granite Mountain, had been assigned to work through the night of June 6. Steed grabbed an information packet that Day's team had compiled about the fire and took his place among the crowd of thirty or forty firefighters. While he waited for the briefing to begin, he flipped through the twenty-four-page packet, which included information on everything from weather forecasts and expected fire behavior to special concerns like the 188 known archaeological sites within the Thompson Ridge area.

The briefing started shortly after 5 P.M. with the incident meteorologist delivering the forecast for that night. These specially trained weathermen use a network of sensors and firefighter-provided information to create forecasts specifically for the fire site. That day, strong afternoon winds had exceeded the meteorologist's expectations and pushed the blaze's eastern front—the pincers—down the back side of Redondo Peak and perilously close to the historic cabins in the Valles Caldera. Day's team had always expected the fire to move east, following the terrain and the prevailing winds, but what surprised them was how quickly the flames had closed the distance on the homes. The strong winds blew the fire through almost four thousand acres of drought-cured trees, doubling the blaze's size in less than twenty-four hours. The forecaster predicted the winds would die down shortly after sunset, which set up a window for the firefighters to go in and protect the cabins.

Darrell Willis, the City of Prescott's Wildland Division chief and Granite Mountain's boss, happened to work for Day's team. On Thompson Ridge, his role was night-operations chief, the tactical adviser who executed the big-picture strategy for how to attack the fire at night. Willis shook Steed's and the squad bosses' hands, but he had more pressing business to attend to. The forecaster handed him the microphone, and Willis briefed the gathered firefighters on the night's actions.

The management team had split the fire into geographic areas. The regions, or divisions, are named alphabetically, with the first, usually the one closest to the fire's head, dubbed Alpha (for A) and the second called Zulu (for Z). The scale leaves room for twenty-four more division breaks, should the fire continue to grow in size and complexity. There were six divisions on Thompson Ridge, and commanding each was a division chief who acted as the field general for his or her piece of the fire. That night, Willis assigned Granite Mountain, along with three other hotshot crews and seven engines, to work under Allen Farnsworth, a retired Bureau of Land Management firefighter from Durango, Colorado, who was in charge of Division Zulu. Willis had tasked Farnsworth with controlling the head of the blaze, and he wanted Steed and the hotshots to try saving the houses by burning the forest.

———

It was almost 9 P.M. when Granite Mountain entered the grasslands on the western side of the Valles Caldera. The hotshots exited the buggies beside a group of hewn-timber cabins. The scenery made it obvious why the Valles Caldera has been used to portray scenes of idyllic frontier living in Tommy Lee Jones, Pierce Brosnan, and Johnny Depp films. The rounded summit of Redondo Peak served as the backdrop. Pines surrounded the meadow. A mountain stream flowed through the caldera. On the night of June 4, most of it was on fire. Darkness seemed to magnify the flames, silhouetting the cabins and the men and women protecting them. It was a classic hotshot fire: wild, dynamic, complex. Renan knew enough to act like he'd been there before. He packed another pinch of tobacco into his dry mouth and held the can out to Grant, who took it readily.

"This is it, boys," Scott said, doing that odd dance young men sometimes do when excited. He bounced on his toes with his arms straight down as if he were holding buckets of water. "This is what we've been training for," he said. "This is the *big* show."

A four-wheel-drive fire engine with extra-high clearance rumbled past toward a stand of torching trees, and a wave of embers blew over a cabin on the edge of the meadow. Another engine crew, emergency lights flashing, cooled the sparks.

When another stand of trees got torched nearby, Grant turned to Renan and actually cupped his ears. "Holy shit!" he yelled. "That's loud!"

There was little doubt that Granite Mountain would be working sixteen-hour shifts until Thompson Ridge was wrapped up. That might take weeks. Physically and mentally, Grant convinced himself he could handle the discomfort, but the promise of so much hard work elicited a feeling closer to dread.

Even standing before such a raw display of nature's power, Grant, like most hotshots, didn't consider his job unnecessarily high-risk. The flames were a source of awe—not fear.

"What are the odds of me dying in a fire?" he'd once asked his mother, to reassure her of his safety. "Think about it, Mom."

Grant's instincts were well founded. In spite of firefighting's apparent dangers, compared with many other professions it's relatively safe. Since fire agencies began keeping track of deaths in 1910, 1,075 firefighters have died on the line, a rate of about 10.5 per year. Between 2000 and 2013, 261 firefighters were killed in the line of duty—considering the fifty-six thousand men and women who are estimated to work fires every year, this is an average ratio of one death for every three thousand workers. For reference, in 2012, an average year in terms of fire-line deaths, the job was considerably lower risk than logging, commercial fishing, roofing, or garbage collection.

When considering whether to become a hotshot, Grant, of course, didn't delve into the relative death rates of firefighters versus trash collectors. His calculus was far simpler: To survive, he needed to trust Steed and the rest of the overhead. In many ways, the relationship between superintendents and their firefighters is most similar to that between mountain guides and their clients. On a well-functioning crew, rookies warn their supervisors of hazards but must trust their leaders to interpret the risks and guide them safely through a blaze. For this reason, "I'd follow them blindly" is a widespread sentiment among firefighters, and among hotshots, "Keep your head down and shut up" is spoken as a mantra. Standing before the hundred-foot flames leaping from the forest, Grant was more than willing to do exactly as Steed told them.

Steed told him to get the drip torches. To meet the fire on their own terms, Granite Mountain and four other crews would burn out along a road that ran between the cabins and the wildfire. The majority of the hotshots would hold the line, watching the cabin side of the dirt road for spot fires, while Bob, Grant, and a few other hotshots worked the torches.

In a staggered pattern, Grant and the other burners widened the line by laying strips of fire parallel with the road. Bob went first, swinging his torch back and forth with the easy swishing motion of a horse's tail. A line of flames spun out behind him. Grant followed a few feet from Bob. He tried to mimic his cousin's fluid movements, but a few minutes into the operation, his wick seemed to clog. The fire stopped pouring out. He stood in a field of quickly spreading two-foot flames and shook the torch back and forth. Still nothing. He pointed the torch upward to check the wick, which further agitated the fuel mix in the steel canister and sent a burst of flame at his face. Before Grant could jerk the torch downward, the flames burst so close to his face that his eyebrows were nearly singed.

"It's fine, Grant!" came a cautionary yell from one of the veterans. The reassurance shook Grant from his moment of stupor. The burst of flames seemed to have cleaned the wick, and when the fuel started flowing again, Grant tripped as he rushed to catch up to Bob.

As Grant burned, the other hotshots spread out along the road, with ten to fifteen feet between each man, and watched for spot fires across the line. Every so often, when a thicket of young pines burst into flames, one of the squad bosses yelled, "Eyes in the green!" For the jittery rookies, like Kevin Woyjeck, ignoring the fireworks was almost impossible. He paced about the road, glancing neurotically from the flames to the unburned meadow.

When a sound like a tropical squall bouncing off broad leaves burst from the forest, Woyjeck couldn't resist the urge to watch the spectacle. The flames climbed into the upper branches of a fir tree and ripped well above the forest's crowns. Embers swarmed upward in the smoke and drifted slowly back to the ground in wide and shimmering arcs. Woyjeck could feel the heat from seventy feet away. Behind the burning fir he saw, on the distant hillsides where the fire had burned hours before, thousands of embers still glowing and pulsing, as if the stars had been tinted orange and reflected off the blackness of the Jemez Mountains. Thompson Ridge was terrifying and beautiful like few things Woyjeck had seen.

"Turby?" Woyjeck said to Bravo's lead firefighter. The erupting tree concerned the rookie.

"Yo," Travis Turbyfill said, his typical monosyllabic response to Woyjeck's questions. Turby, a father of two little girls, had served in Afghanistan as a Marine before coming to Granite Mountain, where he'd worked for two seasons.

"Should we go stand by the house?" Woyjeck asked.

Turby paused and looked at the rookie, as if surprised by the question. He dropped his voice an octave, jutted out his chin, and said, "Nah, let's just stay where we're at, spread out, and keep our eyes in the green."

Turby wanted Woyjeck to stop watching the fireworks and keep his eyes toward the green. A spot fire could threaten not only the buildings but also the men's safety. For the rest of the night, Granite Mountain secured the fire around a half dozen homes tucked into a stand of old-growth ponderosas. Sometime after 2 A.M., the humidity climbed, the fire calmed down, and fatigue began to override adrenaline.

To fight off drowsiness, Zup, Bunch, and his swamper Wade Parker met together on the fire line and took shots of instant coffee at the top of every hour. Zup and Bunch both carried pint-glass-size backpacking stoves in their packs, but that night, they didn't have time to boil water. Instead they shook the packets of grounds straight into their mouths and washed the coffee down with a swallow of water. After their fourth or fifth caffeine cocktail, and almost thirty hours without sleep, delirium took a firm hold.

During one of their meetings, Renan walked up to borrow Parker's tool, and the swamper started chuckling for no reason other than exhaustion. It grew into the sort of insane laughter that feeds on its own sound, and Zup and Bunch were soon infected. With Renan staring at them, the men from Bravo shook with hysterical laughter. Bunch was literally on his back, howling.

Renan, doubtful that anything could be that funny, found it mildly irritating. He had no idea what had made them lose sanity. He snatched Parker's tool and walked back toward the house he was working beside, their laughter pealing behind. In the darkness, Renan moved slowly down the middle of the thin road as the shadows of the flames danced on the pines and white smoke drifted across the road. Suddenly, beyond the crew's clamor and in his momentary silence, Renan felt more exhausted than he had in a very long time.

CHAPTER 11

TESTED

One of the reasons there are so many excellent photographs of the country's wildfires is that land management agencies often hire photographers to document the action. On June 4, the Valles Caldera National Preserve hired Kristen Honig, a thirty-three-year-old planner at Los Alamos National Lab and a semi-professional photographer. The Valles Caldera was happy to have her. Honig had once been a firefighter herself, but even with her experience, Day's management team wouldn't let her onto the line until the fire behavior calmed down. Around midnight, it did, and Division Zulu assigned Clayton Whitted, Alpha's squad boss, to be her minder.

Clayton had accepted a temporary detail as a task-force leader trainee, a position that helped Farnsworth—Division Zulu—orchestrate the movement of resources assigned to his section of the fire. Hotshot crews regularly offer up their most skilled firefighters when needed, and though Granite Mountain was already down a few key overhead, Steed let Clayton go for a few shifts. He trusted that Bob Caldwell, Travis Carter, and Travis Turbyfill could fill the void. The detail would provide Clayton with valuable experience.

Honig was pleased to have a chance to shoot such an active fire.

She sat in the passenger seat beside Clayton, who had parked at the preserve's entrance, where he was directing incoming resources to their assignments. From across the meadow, Honig snapped photos of the fire lighting the sky like a distant city. Clayton wasn't paying attention to her. He was leaning out the window, briefing one of the many resources rushing to the fire. Day had ordered 1,092 firefighters to be on scene in the next few days, and the influx of men and machines had been more or less continuous that night.

"There's a bunch of little marshes and shit out there," Clayton said to one engine captain. "So it's not a good idea to drive the engine off the road."

Clay, who was balding, with a thick handlebar mustache, sent the captain and his engine rumbling down the road toward an old sheep corral, then popped his truck into gear and followed the emergency lights toward the flames.

"Here we go," he said to Honig. "Now, let's see about getting you some good pictures, then."

Crews had started the firing operation on the east side of the caldera, and the burnout was moving back toward the wildfire in the west. As they moved along the fire's edge, Clayton stopped often to check in with firefighters spread out along the road.

The conversation consisted mostly of Clayton deflecting requests for additional resources and more water. The firefighters were stretched uncomfortably thin, and though the burnout was holding, Clayton and everybody else in Division Zulu was aware that one strong gust of wind could change everything.

While he worked, Clayton did what he could to help Honig get good photos. At one point, the fire intensity picked up in the meadow, and he led her on foot toward the fire's edge to get close-ups. Then the breeze shifted, and both hustled back to the truck through blowing smoke. "You doing okay?" Clayton yelled to her. She couldn't see much beyond the glow stick that Clay had attached to his helmet— a safety precaution he and the other hotshots had taken to make themselves more visible. But yes, Honig said, she was doing just fine.

When they met up with Granite Mountain, Grant and Bob were still burning out. Clayton called Bunch and Turbyfill over and asked them to pose for a few pictures.

"You're not with the news, are you?" Bunch asked Honig, bristling a little at the sight of a camera. "'Cause we had a lady once who followed us around for days." Then he mumbled, "It got a little weird."

"What's that?" Honig hadn't heard him.

"It got a little weird," he said, this time clearly, if not too loudly. "She was at spike camp and stuff."

"Yeah?" Honig asked. She wasn't quite sure how to respond to this comment, and Clayton recognized the awkwardness and politely stepped in to explain who Honig was, what she was doing, and why Granite Mountain was happy to help. But what struck her more than Bunch's odd story about the other photographer was the hotshots' utter lack of concern. The wildfire was feet away from the men, the house, and its 250-gallon propane tank, and the entire forest was glowing as if lit by floodlights, but as they worked, the hotshots' primary concern seemed to be tobacco. They called it bung.

"Turby, you want some bung?" Bunch yelled to him. The normalcy of their conversation put Honig at ease. She asked Turby to pose near the cabin, framing a shot of the big ex-Marine silhouetted by the fire. Clayton looked over her shoulder.

"Well, I wanted to be your main subject," Clayton told her, his voice hushed and fringed with western twang. She laughed.

"But I don't have no gloves on," he said.

He was referencing a Fire Service edict that no photos or video are to be shared unless all rules—even those often broken, like wearing gloves—are being followed in the images. Honig liked Clayton immediately.

At twenty-seven, Clayton was in general either unconcerned or unaware of what people thought of him. That quirky confidence was part of his charm. He'd made a hobby of thrift-store shopping and used Granite Mountain's station, where he worked all year as one of the hotshots' six permanent crew members, as a gallery for his collection of useless kitsch: a strip of Halloween paper skeletons, a stitched velvet picture of Jesus, an antique chainsaw. Naturally easygoing, Clayton tried not to take firefighting—or life—too seriously. Years earlier, during a nighttime burning operation similar to the one on Thompson Ridge, a friend of his had spent his birthday on the fire line. Clayton, who was working on the other side of a lake, grabbed a drip torch and burned his friend's initials—W.W.—into the side of a mountain. W.W. could see it from a half-mile away.

Honig and Clayton spent the night working together. By morning, she needed to return home to Los Alamos but found that she didn't want to go. She didn't know why, exactly, but she was enjoying herself. By dawn it was clear that the operation had been a success. The burnout on Division Zulu had blackened the fuel around the cabin. No homes had been lost, and now, with cooler nighttime temperatures and higher humidity, the fire gathered itself in the heavier logs strewn across the forest floor. It wouldn't become dangerous again until the day got hot, that afternoon. The day shift could deal with the fire's reawakening. For now, the firefighters on Division Zulu could pause.

Clayton parked the truck facing east, and he and Honig sat in the cab watching the sunrise through the low smoke that hung above the caldera like morning fog. Elk herds with young calves grazed in the forest just beyond the creeping flames, and coyotes and bears sporadically passed through the meadow. A few days later, firefighters would rescue a burned cub and give it the name Redondo. As they sat in the cab, Clayton handed Honig his iPod and asked her to pick out the music.

"Much of it is my wife's," he said, adding, "She's a second-grade teacher."

"Whatever I pick, you're going to hate," Honig warned. She only listened to country and Christian music, and few firefighters she knew listened to country, and even fewer to Christian. But she scrolled past his archival selection of Tupac albums and recognized in the hip-hop collection Casting Crowns, a Christian rock band.

"I love 'Set Me Free,'" she said, a little self-consciously. It was one of Casting Crowns' more popular songs.

"Me, too," Clayton said. In a field as rough as firefighting, conversations about church and faith are rare.

Clayton told Honig that the job actually brought him closer to Christ. Six years earlier, he was working with the Prescott Hotshots when his mom died of a brain tumor—a similar illness to the one that had stricken Chris MacKenzie's mom while Granite Mountain was on the Hart Fire that week. As Chris had done, Clayton left the Prescott Hotshots in the midst of fire season to help take care of his mom. During her sickness, he became a youth pastor at the Heights Church, and she died in 2008, a year after getting sick. Clayton returned to the fire line the following spring.

He found that hotshotting and ministry work had their similarities. Both attracted followers who were young and drifting. Firefighting, though, was his calling, Clayton said. That was part of the reason he'd come back to the line. It gave him an opportunity to share his faith in a subtle way.

Before he dropped Honig off at her car, she told him that her father had recently died of cancer. His decline had taken the better part of a decade.

"Do you think it's easier for the family if the death is long and drawn-out?" Honig asked, and she looked at Clayton, his face streaked with ash nearly as dark as his mustache. "Or is it easier if a loved one goes quickly?"

Every season, there's one particular fire that teaches rookies what it means to be a hotshot—why the job is so self-selecting, why hazing exists, why the physical training is so demanding. Enduring these fires gives most young hotshots an unassailable pride. The intensity of the experience also breeds a shared mindset: If you've been there, you understand the commitment the job demands; if you haven't, it's something that's difficult to fathom. As hard as these fires are, they also forge friendships. For Grant and Renan, their test was Thompson Ridge.

After the third night shift and their seventh day since leaving

Prescott, the battle buddies waited by a barn in the caldera's meadow for a displeasing breakfast of dehydrated eggs. For the first time in days, they were far enough from the other hotshots that they had space to speak openly.

"How you doing, man?" Renan asked Grant.

"I'm struggling," he said.

Grant was tired of being watched and judged; tired of the grime, the smoke, the bad food, the lack of sleep. Just tired. The work was mostly burning and holding, not particularly hard, as hotshotting goes, but since the first night shift, Donut was the only person on the crew who had slept much or well. He'd brought Tylenol PM and had the foresight to take a healthy dose before lying down. The rest of the crew suffered. They slept in a campground a fifteen-minute drive from their section of line, and during the day temperatures often climbed into the nineties. For most of the eight hours Grant and Renan were supposed to be sleeping, they lay on their mats swatting flies and sweating. By 5 P.M., when Granite Mountain reported to fire camp for their night shift, they'd slept for three hours, maybe four.

"I've been thinking a lot about home," Grant said. "About Leah." The last time he'd talked to her was after his spat with Tony on the Hart Fire, and the work—and his loneliness—had only gotten harder since then. He told Renan how he'd fallen for Leah.

Four days after they'd first met, Grant called and asked Leah if she wanted to go to the beach. She'd never been.

"When?" she asked.

"Now," he said.

Eight hours later, they were in Newport Beach, California. Grant and Leah met Bob, his cousin, at their grandmother's house, and immediately after, the trio went to a restaurant at the end of the Santa Monica Pier, then to Venice Beach and to Hollywood.

Later that night, Grant led them to the empty Disneyland parking lot. It was late night by now, and he parked between streetlights and put on a hip-hop album that Bob had bought from a musician on the street that morning. Grant and Leah held hands in the front seat, Bob drank canned Coors beer and got funnier in the backseat, and all three of them watched the Happiest Place on Earth glow in the darkness.

"You're coming to our wedding," Grant said to Renan in the meadow. He plucked a blade of grass and folded and tore it into neat strips. Grant had known Bob his whole life, but his cousin had other responsibilities on the line. Renan understood what Grant was going through better than anybody else.

Renan felt the same. They were partners who literally spent every waking moment together. As battle buddies they had to. They also wanted to. The newness of the fire line made them crave a companion to share and contextualize the experiences. That morning, Renan told Grant about his girlfriend and the ring he'd bought six months earlier. He wasn't going to give it to her until he had a structural firefighting job and enough money to pay for the wedding.

"You and Leah will be there when that happens," Renan said.

Scott walked up, and the rookies put their conversation on hold. He lingered long enough that Grant volunteered, "I'm feeling it, dude. I'm worn out, I'm missing home—my dog, my girl."

"Oh, little rookie's missing his girl?" Scott asked. Renan prepared for the belittling and inevitable jokes that often pass for conversation among young men. But they didn't come. Scott realized that now wasn't the time. He changed tack.

"It's your first fire. Everybody wonders what the hell it is they're doing out here," Scott said. "That's normal. It's a phase, though."

Scott had plenty of slow moments on Thompson Ridge in which to worry about the puppy and unpaid bills, fantasize of past lovers, create and dispel fears of losing Heather again. He also thought about his sister. She was due any day. He wanted to be there for his nephew's birth.

"It gets easier," Scott said to Grant, perhaps with more conviction than he felt.

"Do you remember your first hard shift on a fire?" Grant asked.

"Oh, man, of course I do," Scott said. But he didn't tell them about it. Instead he just offered some simple advice. "There will always be times when you get sick of this shit. Having somebody at home makes it harder. It makes it feel like you're never fully here," he said, squeezing Grant's shoulder. "Just grit your teeth and you'll get through it."

They could see a couple of green Forest Service pickups bouncing down the road toward the barn, dust billowing behind them.

"You're doing good, dude," Scott said.

When the trucks pulled up next to the barn, the men unloaded white buckets filled with hot plates of breakfast. Grant, Renan, and Scott lined up along with the other crews spooling out from the folding tables and rejoined the world they'd momentarily left. After breakfast, fed and exhausted, the men in Alpha buggy returned to the campsites in silence. There was no music, and each spent the short drive alone with his thoughts. Whatever euphoria had driven Bunch, Zup, and Parker to momentary madness had evaporated, and a deep fatigue now loomed over the crew like a great collective hangover.

After unloading the other hotshots' bags, as they always did, Renan and Grant slept next to each other. They unrolled their pads and sleeping bags in the shade of a pine that great age had given character. Grant neatly stacked and folded his ash-covered pants and shirt beside his bed. Renan sighed when he took off his boots and sat for a long moment rubbing his naked and shriveled feet. Colonies of athlete's foot were now well established in his toes, and above his sock line was a white ring that separated the dirt above his calf from the damp white flesh below. His body ached. Letting his feet breathe was a small luxury.

As the other hotshots collapsed into their bags, Grant pulled out the tobacco he'd bought and took a seat on a picnic table nearby. Finally alone, he rolled a cigarette. Since the gas station, smoking had become his quiet ritual—a small creature comfort before bed. Watching Grant smoke was the last thing Renan remembered before falling asleep.

Renan woke with a sharp inhale. He knew this feeling. He hated this feeling. His back was seizing. Renan grabbed Grant, who was asleep beside him, and pulled him closer. Grant was slow to wake and turned to see Renan's eyes wide with panic as he slipped from consciousness. Renan began to writhe. Grant, now aware that something was very wrong, put his face close to Renan's and screamed.

"Renan! Renan!"

His shout roused the rest of the hotshots. Somebody called an ambulance. Grant was kneeling over Renan when he regained consciousness. Renan saw only his friend's face hovering six inches above his own. Grant kept repeating Renan's name, and it sounded hollow and meaningless, but even through the lens of shock, he could tell Grant was terrified.

Renan's body convulsed and blackness returned. Grant put a hand on his friend's forehead and the other on his arm, and the tears streamed down as he felt Renan arch and quiver and arch and quiver. The entire crew surrounded Renan in a wide circle. Most were shirtless and in their boxers, trading guesses in hushed tones about what was happening to Renan and ideas about how to keep him safe.

When consciousness returned, it came in pieces. Hearing was first. Renan didn't know who was speaking or what they were saying. The paramedics, who arrived within twenty minutes, loaded him onto a gurney and spread a wool blanket over him. The hotshots parted as he was wheeled to the ambulance. The first thing Renan remembered clearly was Grant and Donut framed by the ambulance's side door.

Grant was crying. Donut had been. He grabbed Renan's motionless hand and squeezed.

"You're going to be okay, man," Donut said. "We're here for you."

Tears welled in Renan's eyes. He knew he wasn't coming back to the crew. The paramedics closed the back door and said it was time to go. Grant looked down and tried to find the right words to reassure his friend but failed. The door was shut and Grant, Donut, and the rest of the crew watched the ambulance pull away.

CHAPTER 12

WE'RE STILL HERE

Tragedies both large and small punctuate the history of America's long war on wildfire. In 1871, fifteen hundred people died in Peshtigo, Wisconsin, a logging town near Lake Michigan, when, according to legend, a flaming front five miles wide and twice as tall as the Washington Monument swept through town. That same day, five hundred more people died in two separate large fires in Michigan and Illinois. By the end of the nineteenth century, wildfires would kill 836 more Americans, most of them civilians.

But for all the casualties, no tragedy has shaped America's fire policy more than Idaho's Big Burn. The same historic firestorm that made Ed Pulaski famous for saving forty-five of his crew members' lives went on to kill eighty-seven people and blacken three million acres. Far from America's largest or deadliest blaze, the Big Burn's significance lay in its timing.

The fire sparked in August 1910, just five years after Theodore Roosevelt created the Forest Service and placed under the new agency's control an area twice the size of Montana. The conservative Congress, many of its members deep in the pockets of timber barons, was incensed. Roosevelt had used an executive action to bring millions of previously unclaimed acres under federal control, and he did so, in his typically willful style, without consulting Congress.

Congress, which sets the annual budgets for federal agencies, responded to Roosevelt's brash action by giving the Forest Service an anemic and unsustainable budget. Two of the agency's early directors resigned in protest of their low salaries, but it was even worse for the rangers, who received only $900 per year and were required to provide their own horses and saddles.

If the Forest Service was going to survive beyond its infancy, the agency and its charismatic first director, Gifford Pinchot, needed more rangers and more money to manage the land. They needed to move beyond the political battles and convince Congress that their relevance was greater than merely being caretakers of the wilderness. The key, once again, lay with the politicians beholden to the timber industry, which saw every blackened tree as a lost dollar. Pinchot and the leaders of the Forest Service quickly realized that controlling wildfires would become the agency's way to secure sufficient funding.

The Big Burn made the agency's case. Three million acres of forest burned in one mammoth firestorm, and the timber industry calculated the losses to be in the millions. Pinchot's Forest Service rangers, though, were the ones leading the effort to extinguish the flames. They organized civilian militias raised from bars and churches and helped direct the Buffalo Soldiers and other cavalry when the Army got involved. In the Big Burn's aftermath, Congress finally found purpose in the agency. Their job, as one early Forest Service director put it, was to keep the land ripe for the ax by suppressing wildfires.

Over the decades that followed, the Forest Service and the timber industry grew to depend on each other. Loggers clear-cut national forests in every western state, and the agency grew wealthy from the leases the companies paid to harvest publicly owned trees. To better curb the only scourge slowing the timber harvests, the agency invested heavily in fighting fires. During World War II, conscientious objectors served their country by fighting wildfires. The Forest Service shortened response times to wilderness blazes by building roads into remote mountain ranges, erecting watchtowers among swaths of virgin forest, and founding the smoke-jumping program—an idea borrowed from Army paratroopers—to extinguish any fire too remote to hike or drive to. Army surpluses left behind after World War II—Jeeps, air tankers, helicopters—further militarized suppression troops, and the public was enlisted in the war on fire.

During World War II, the government issued posters that showed Hitler and an Imperial Japanese soldier grinning beside a wildfire. The Japanese had taken to sending high-altitude hot-air balloons across the Pacific to set ablaze America's forests and its cities. They launched an estimated nine thousand balloons, and though 342 reached the United States, only one caused significant damage. The hydrogen-filled rubberized silk balloon landed in southern Oregon, killing a pregnant Sunday school teacher and five of her teenage students. They were the only U.S. combat casualties in the forty-eight states. The text beneath the government-issued campaign posters read, OUR CARELESSNESS, THEIR SECRET WEAPON.

Ultimately, though, it was a shirtless, shovel-wielding, ranger-hat-and-blue-jeans-wearing brown bear that colored America's perception of fires. Created by the Advertising Council in 1944, Smokey Bear tapped into the nationalist fervor that had swept the country during the war years, and through radio and TV ads, folk songs by popular country musicians, and movies, like *Bambi,* that villainized wildfire, the phrase "Only You Can Prevent Forest Fires" became ingrained in the American psyche. The advertising campaign became one of the most successful messaging efforts in history. By the twenty-first century, Smokey's fame rivaled that of Santa Claus. At a glance, 95 percent of adults and 77 percent of children knew Smokey Bear and his anti-wildfire message.

Many firefighters were dying trying to corral the burns. In the first thirty years of the Forest Service's grand experiment to control the flames, 186 firefighters were killed on the line. These fatalities only strengthened the agency's resolve to suppress fires. Smoke jumpers, fire engines, air tankers, helicopters, and hotshot crews were developed to aid the effort to extinguish any and all sparks in the forests.

Before suppression became the reigning management policy in the

1930s and '40s, forest scientists estimate that more than thirty million acres burned in the West every year; after suppression was instituted, most fire seasons were held to less than five million acres. By the mid-1950s, firefighters were stopping almost all wildfires before they reached the size of a city block. In just forty years, the Forest Service had assembled the world's most effective firefighting force.

What forest managers hadn't foreseen was suppression's serious environmental consequences. Unwittingly, land managers had made more volatile. Naturally and regularly occurring fires played a crucial role in the evolution of most western forests. Removing the flames changed the self-regulating system. As a strategy for dealing with natural wildfires, dousing every spark worked just long enough for countless pine, fir, and cedar saplings to grow into mature trees. By the end of the twentieth century, the policy failures were nowhere more obvious than in the Southwest.

Suppression had changed the forests. The Southwest used to simmer with frequent low-intensity burns every summer. Lightning ignited some blazes, but native peoples intentionally set the majority to cultivate grass for the elk and deer herds they relied on for food. Smoky summer skies were the norm. Towering columns associated with extremely hot blazes were not. Fires burned at low intensity, with flames not much taller than knee height. These blazes meandered across the landscape, clearing the forest of underbrush every ten to fifteen years. Left behind were open meadows and ponderosa forests with widely spaced canopies and black charcoal scarring the trees' trunks.

Without frequent blazes, brush and saplings choked out the open grasslands and created a ladder of fuels that grew from the forest floor to the tops of the pines. Since suppression began, millions—maybe billions—more trees have crowded into western forests. Though firefighters have limited the frequency with which fires escape, the ones that do now have exponentially more fuel to burn. In one New Mexico mountain range, there are now 1,300 trees per acre where a century earlier there were just 150. Flames can climb quickly up the ladder and into the crown, where fires ignite the tallest tree and create infernos that are nearly impossible to control.

Nationwide, firefighters began noticing an uptick in fire intensity in the early 1980s. But it wasn't until the summer of 1988, when more than a third of Yellowstone National Park burned, that the greater public took note of just how unhealthy the forests had become. For three months the fires raged, and national newspapers splashed across their front pages images of two-hundred-foot flames dwarfing park buildings. Dubbed the Summer of Fire, the conflagration closed the park to tourists, and Americans were concerned that they were witnessing the destruction of a national treasure.

"People were horrified by Yellowstone," said Harry Croft, the deputy national director of fire and aviation management for the Forest Service during the mid- to late seventies. "We didn't like what was going on, either. So we put together a plan that would reintroduce fire, logging, thinning—something to get our forests back on track and limit how often we get the Yellowstone anomalies."

Croft and his colleagues released an updated National Fire Plan in 2000. It allowed fires that didn't threaten houses, endangered species, or watersheds to burn unmolested by firefighters. The policy shift is part of the reason six-million-acre fire seasons are the new norm, and ten-million-acre seasons loom on the horizon.

"More acreage burned is a very good thing," says Alexander Evans, a forest scientist who studies wildfires for the Forest Guild, a Santa Fe–based nonprofit that advocates for sustainable forestry. Before World War II, fire seasons that burned between thirty and forty million acres, much of it at a lower intensity, weren't uncommon. "In a few places, we're even seeing a return to natural fire cycles and the fuels getting thinner. That's what we want more of," says Evans.

But now that we've made it possible for forests to grow unchecked for years, it's not as simple as it sounds. As acclaimed fire scientist

Stephen Pyne puts it, removing fire from the landscape is much easier than putting it back.

The combination of dense forests and drying climate has made highly destructive fires—like the Yellowstone anomalies—increasingly common. Arizona, New Mexico, California, Nevada, and Colorado have all seen their most destructive blazes since the turn of the twenty-first century.

Forest scientists and the media alike have branded these blazes mega-fires. The term is imprecise and means only that a burn is destructive, extremely difficult to control, and, because of this, expensive. The term's usage also turns a blind eye to the fact that large wildfires, though somewhat anomalous, have historically shaped American forests. Between 1825 and 1910, six documented fires—in present-day Wisconsin, Michigan, Maine, Idaho, and South Carolina—exceeded one million acres. Each of them burned more acreage than the largest blazes the United States has seen in a hundred years, and three of them were larger by a factor of five.

What the term "mega-fire" accurately reflects is that fires are burning more intensely now than at any point since the Forest Service became proficient at the business of suppression, and land management agencies are having a hard time coping. Though the Forest Service,

Historic and Recent Significant Southwest Fires

1910	1949	1966	1988	1990	1994	2000
GREAT IDAHO/ BIG BURN OF 1910	MANN GULCH	LOOP	YELLOW-STONE	DUDE	SOUTH CANYON	CERRO GRANDE
LOCATION: **Idaho and Montana**	LOCATION: **Montana**	LOCATION: **California**	LOCATION: **Wyoming**	LOCATION: **Arizona**	LOCATION: **Colorado**	LOCATION: **New Mexico**
Multiple blazes burn 3 million acres and kill 85 firefighters in a matter of days. The start of America's war on fire.	13 smoke jumpers killed.	13 hotshots killed.	Multiple blazes burn 1.2 million acres; media keys into the deteriorating health of America's forests.	6 firefighters from inmate crew killed.	Fourteen elite back-country firefighters die.	Escaped prescribed burn becomes most destructive in U.S. history, causing $1 billion in damages.

the BLM, and state agencies are still aggressively trying to control mega-fires, there's a mounting body of evidence suggesting that it's a costly and failing endeavor. Jerry Williams, the former national director of fire and aviation management for the Forest Service, who is credited with popularizing the term "mega-fire" in the mid-2000s, estimates that only one-tenth of 1 percent of the fires that burn each year qualify as mega-fires. Yet the severest 1 percent of blazes account for 85 percent of the suppression costs—hence the upper echelon's significance.

"The worst wildfires on record are coinciding with a time when suppression force has never been greater, technological advantage has never been better, and suppression spending has never been higher," Williams wrote. "Mega-fires challenge the commonly held notion that increasing wildfire threats can be effectively matched with greater suppression forces."

Williams's report isn't the only one to put forward these views. In the past few years, one Forest Service–funded report suggested that fully three-quarters of air-tanker drops are ineffective during the initial attack phase of the firefight. Rapidly spreading fires burn straight through the slurry, which is exactly what Todd Lerke saw on Thompson Ridge. Yet because the tactic is widely and perhaps erro-

2011	2011	2012	2012	2012	2013	2013
LAS CONCHAS	WALLOW	WALDO CANYON	WHITEWATER-BALDY	LITTLE BEAR	BLACK FOREST	YARNELL HILL
LOCATION: **New Mexico**	LOCATION: **Arizona**	LOCATION: **Colorado**	LOCATION: **New Mexico**	LOCATION: **New Mexico**	LOCATION: **Colorado**	LOCATION: **Arizona**
Largest fire in state history threatens nuclear waste dump at Los Alamos National Laboratory.	Sets record for largest fire recorded in the lower 48, burns 538,000 acres.	Kills two civilians and burns 347 homes, sets record for most destructive fire in state history.	Sets new record for largest fire in state history, burning 297,845 acres.	Burns 242 homes and sets new record for most destructive fire in state history.	Sets new record for most destructive fire in state history, kills two civilians and burns 511 homes.	19 Granite Mountain Hotshots killed.

NOTE: *Since 2000, the following states have also seen historically destructive fires: Washington, Oregon, California, Idaho, Texas, Georgia, Minnesota, Florida.*

Research by Amanda Eggert

neously recognized as effective, incident commanders drop retardant on the vast majority of quickly spreading blazes. Another report found that the costs of fighting fires fall by half when the new blazes burn into the blackened fuel left behind from previous fire scars. In other words, when *more* fires are allowed to burn, fire size regulates naturally, and controlling blazes becomes *cheaper*. It's likely to take many years for this information, and other studies like these, to drastically reshape the way we deal with wildfires in America, but the fact that critical research is forthcoming is a sign that the Forest Service, the Bureau of Land Management, and state forestry agencies are open to adapting their management policies to the new realities of wildfires. The change hasn't happened yet, and in the meantime, firefighters are still stopping 98 percent of burns in the initial-attack stage.

The eighteen Granite Mountain Hotshots remaining on the Thompson Ridge Fire had little time to process Renan's departure. No one knew whether he would live, or the extent of his injuries if he did, but the men were clearly disturbed by what they'd witnessed. After the ambulance rushed toward an emergency room two hours to the south, they clumped together in knots of twos and threes before pulling apart and lying back down for another few hours of rest. Grant was inconsolable. He called Leah, sobbing. Bob pulled him aside and told him to settle down.

"I know this is hard, but you've got to get over this," he said.

"We go back to work? That's how we deal with this?" Grant asked.

The simple answer was yes. Granite Mountain still had seven more shifts before Day's team would release them from Thompson Ridge, and the crew's trials were far from over. A few days after Renan's incident, Steed called the men together beside the barn in the meadow. It was after another night shift, and the hotshots, still trying and mostly failing to sleep in the campground, looked exhausted. Ash and charcoal colored their scruffy faces gray, and dried sweat left behind rings of salt on their black T-shirts. They all wanted an update on Renan.

Steed knew only that Renan was going to live, but he didn't know the details about his condition. He did have other news, though, and it wasn't good.

That day, a twenty-eight-year-old smoke jumper in California, Luke Sheehy, had died on the line. He and two other jumpers had parachuted into a single-tree lightning fire in Northern California's South Warner Wilderness. A burning branch had fallen sixty feet and struck Sheehy in the head. Despite his fellow jumpers' attempts to resuscitate him, he died before arriving at the hospital.

Steed and the crew took a moment of silence. The more religious hotshots hung their heads in prayer, and the others just shook their heads. None of them knew Sheehy personally, but the news still hit close to the heart. Sheehy had died doing what Granite Mountain was doing every day on Thompson Ridge: work that wasn't exceptionally hard or patently dangerous. But his death made it difficult to ignore the reality that, in one dark moment, the same fate could befall any of the hotshots.

Granite Mountain loaded into the buggies, and as the men headed back to the campground, they steeled themselves for another shift on the line. By the afternoon of June 8, Thompson Ridge had roughly doubled in size since the hotshots' arrival. The blaze now spread across 18,500 acres and threatened another subdivision on its southern edge. To protect the homes, Day's team called for a second burnout. This time, the objective was a four-thousand-to-five-thousand-acre pocket of pines and aspens adjacent to the homes. If Day's team successfully orchestrated the burnout behind the houses, Thompson Ridge would be nearly contained.

The operation was controversial with the public from the outset. One of the homeowners' major concerns was mudslides. Normally, when rain falls on unburned forest, 98 percent of the water soaks into the ground. But the blast-oven heat of an intense wildfire changes the soil's composition and leaves behind a hydrophobic layer. When the monsoon rains come, the water streaks off this waxy film, carrying with it whatever ash the fire leaves behind. After one particularly intense fire in the Jemez Mountains, the rain from a single thunderstorm moved thousands of tons of ash—five inches deep in some places—off the mountains. As black torrents cascaded down toward the Rio Grande, they picked up boulders and logs, and in the deluge, orchards and farms that had been spared from the flames were destroyed by the floods. The aftermath made it look as if the Jemez Mountains had erupted.

Mudslides were one concern. But for homeowners, the greater issue with Day's burnout plan on Thompson Ridge was its echoes of 2000's historically destructive Cerro Grande Fire. Started as a prescribed burn intended to thin out the increasingly thick forest near Los Alamos, the blaze escaped and led to a nuclear scare when flames jumped onto the National Laboratory's property, where all manner of explosive and radioactive materials are used, tested, and stored.

The fire burned for three months, and firefighters were able to contain Cerro Grande only after the weather changed. By then the blaze had razed 235 Los Alamos homes and caused $1 billion in damages. In its aftermath, the name Cerro Grande became shorthand among both firefighters and the press for an increasingly accepted reality: man's reign of controlling wildfire was quickly coming to an end.

An AStar helicopter took off from Thompson Ridge's fire camp shortly before dark. The helicopter banked over the tops of the pines and flew straight toward a patch of unburned forest that sat between a small group of houses and the wildfire. When the pilot looked down and saw the line of flames advancing through the sea of trees, he signaled to a crew member who sat in the wind near an open side door in the back. Balanced on the edge of the fuselage, extending just over the helicopter's skids, was a waist-high box filled with plastic spheres about the size of Ping-Pong balls. The firefighter flipped a switch on the top of the box, an electric motor whirred to life, and every few seconds a ball injected with a highly flammable combination of chemicals rolled down a tube, dropped from the helicopter, and burst into flames just above the forest floor.

From where Granite Mountain was spread out along a road, holding the burnout operation's southern edge, it looked as if the helicopter were firing tracer bullets into the hillside. It didn't look quite right. The smoke was too black, and the fire much too hot, as though the ignition boss in the helicopter hadn't fully accounted for just how dry the forest was. After seeing the flames, firefighters on scene estimated that rather than dropping six hundred balls, as it was supposed to, the helicopter had dropped six thousand.

Within the hour, thousands of tiny blazes climbed up brush and saplings and into the forest's crown, where the flames caught gusts of wind and grew into one massive front. Like massive ocean waves, flames crashed through the forest. One of the fire's heads raced toward the line where Granite Mountain was holding line. It hit so hard that the flames jumped a thirty-foot-wide road and divided another hotshot crew that was spread along the road uphill of Granite Mountain. The fire cut off a few hotshots from the rest of their crew. They abandoned their heavier gear and rushed west to safety while the rest of their crew fled east into the valley. When Bunch and his swamper, Parker, saw the hotshots racing toward the line, they shared a quick and concerned glance: Thompson Ridge was off and running.

When the temperature inside the smoke column exceeded four hundred degrees, whatever bits of aspen leaves, pinecones, or needles were caught in the rising air combusted and became natural-born incendiaries that rode gusts of wind up and away from the main fire. From the valley floor, the rising embers looked like endless strings of Christmas lights wrapped around a rotating smoke column. The effect was supernatural.

The most remarkable story of a fire's behavior comes from the horrific Peshtigo Fire, a blaze that killed an estimated fifteen hundred people in Wisconsin's North Woods in 1871. On the Peshtigo Fire, the temperature difference between the smoke and the cooler air on the blaze's perimeter was so extreme that a tornado formed in the turbulent interface. The fire spun into a cyclone that rose hundreds of feet above the forest. As it twisted through the logging town of Peshtigo, the tornado incinerated grain elevators and lifted locomotives completely off the railroad tracks. The heat became so intense that silicates that had melted out of the soil were sucked into the air. When the firestorm produced a thunderhead, the silicates became molten glass that rained down from the skies. In the aftermath, a thin sheet of glass encased everything—from the thousands of birds suffocated midflight by the fire's insatiable appetite for oxygen to the fleeing family killed atop their horse-drawn wagon.

On Thompson Ridge, there were no flaming tornadoes, but many of the same conditions were present. Like a volley from archers, embers caught in the smoke were shot well ahead of the main fire. Some fizzled and died in rocks or ponds, but others landed and found life in the fertile bed of cured pine needles. Within a half-hour of the helicopter's burnout, the fire had jumped well beyond its lines and kept running. Commuters could see flames from Albuquerque, seventy miles to the south.

Clayton Whitted, who was still serving as task-force leader, was working closely with Allen Farnsworth, who was in charge of Division Zulu. They were together when the fire took off.

"Sure as shit that's gonna jump the line," Clayton said to Farnsworth.

Shortly after the flames crossed the road, they took stock of the damage and drove up to the saddle where the fire had broken through the lines. Hoses lay smoldering beside a melted gas canister and a few charred boxes of MREs and Gatorade. Plumes of smoke corkscrewed into the sky around the truck, and embers bounced off the cab. There were trees torching and dead trees falling around the saddle.

"What do you think, Clay?" Farnsworth asked.

"I don't think there's anything we can do about this now," Clayton said. "I think we should wait."

"Right answer," Farnsworth said.

But it wasn't the answer the day shift's operations chief, Darrell Willis's counterpart and the architect of the burnout, wanted to hear. The management team had already racked up a cost of more than $9 million, and this preemptive fire was supposed to end Thompson

Ridge—not revive it. If another man-made burn spiraled out of control and wreaked havoc on Los Alamos, and this time because some airborne pyro had mistakenly launched ten times more flaming Ping-Pong balls into the forest than he was supposed to, it would blacken the eye of the federal and state firefighting agencies in the area. Somebody was going to have to fix the mistake immediately. And the only somebodies who could do it were the hotshots.

If Clayton didn't like the assignment, Steed liked it less, especially in light of how smoke jumper Luke Sheehy had been killed. The fire was moving quickly in a patch of dead but still-standing trees that, weakened by the fire and burning, could easily be knocked over in a strong wind. Steed felt that chasing the slop-over that night violated or compromised too many of the Ten and Eighteen:

> In country not seen in daylight.
> Unfamiliar with weather and local factors
>   influencing fire behavior.
> Getting frequent spot fires across line.

If there was any doubt among Granite Mountain's crew members about how their inexperienced superintendent would respond to pressure applied by incident commanders, Steed put them to rest on Thompson Ridge. He wasn't willing to risk his men's lives to chase a dangerous fire that in the current conditions they were unlikely to catch. Even at the behest of an insistent operations chief, Steed flat-out refused the assignment.

"But that's what hotshots do!" the operations chief, a former hotshot himself, told Steed. He wanted Granite Mountain to chase the slop-over before it became a disaster for the town of Los Alamos.

Steed was firm. Granite Mountain wouldn't go. Nor would the other hotshot crew working the night shift. It was too risky. But after declining the assignment, Steed offered a productive alternative: Granite Mountain would go after the slop-over as soon as the fire intensity died down. The operations chief consented and, after midnight, the hotshots lined out behind Steed and he marched his men toward the flames.

Born and raised an hour northeast of Prescott in the small town of Cottonwood, Steed lost his mom to a car accident when he was twelve. He joined the Marine Corps after high school and served as a gunner for four years. Though he never saw combat (he was out of the service before 9/11), his time in the Marines shaped his perspective on fire-fighting. "It's the next best thing to the military," Steed often said.

He started his fire career with the Forest Service's Prescott Hot-shots, a crew based across town from Granite Mountain, when he was twenty-five. Over the next eight seasons, he worked on Forest Service engines and a helicopter, where he generated a collection of superhu-man stories firefighters would tell long after Steed left each crew.

He once propped up a falling tree by throwing his shoulder into the trunk, and in a blaze in Yosemite National Park, he rappelled from a helicopter to put out a number of purported spot fires that had been picked up by a thermal imaging sensor. The red dots—heat sig-natures on the computer screen—turned out to be black bears. When Steed touched down, he unclipped from the rope and chased the bears off with his chainsaw.

Over a dozen years of firefighting, Steed's reviews were unwaver-ingly exemplary: "ensured safety and morale during a 32-hour shift," "always weighs safety first," "an excellent listener and very much a people person." By the time he came to Granite Mountain in 2009, Steed had developed a reputation as a composed and intensely fit fire-fighter. He moved quickly through the hotshots' ranks, and after just two years as a squad boss, Steed was Marsh's first choice for captain when the position opened up. "The missing piece of the puzzle," Marsh called Steed.

The prospect of permanently taking over as superintendent thrilled Steed. Already he called Granite Mountain "my crew" and was changing the crew's culture through a leadership style that was starkly different from Marsh's. If Marsh was the stern father figure,

Steed was the cool older brother. With a picture-perfect family—a pretty wife, Desiree, a four-year-old boy, Caden, and a three-year-old girl, Cambria—he had always been a natural role model for Granite Mountain's more family- and community-oriented hotshots. But when he took over as superintendent, his charisma and physical prowess served to make him a role model for most of the men. Steed reinforced that status by often telling the hotshots he loved them. He also did away with some of the crew's less popular protocols.

He relaxed the dress code, allowed the men to wear their beards thick, and introduced new traditions beyond just the intensity of the physical training. As punishment for a poorly mopped-up log or a fuel bottle accidentally left behind, he'd forced the whole crew to drop down in the ash and lead them through forty push-ups. But more often he rewarded them, allowing the hotshots to play Frisbee in the afternoons or springing on them impromptu team-building exercises like building a human pyramid. Every Saturday afternoon, during the hotshots' work hours, Steed invited the men's families to come to the station to barbecue. By May, the picnics were robust affairs. Wives, girlfriends, and parents all came. One Saturday, a troop of more than a dozen kids traipsed about the station.

Donut and a few of the other veteran hotshots sometimes equated Steed's relaxed approach to leadership with a tolerance for laziness. Bunch didn't, though. He thought Granite Mountain was in both the best shape and the highest spirits he'd ever seen.

The crew's lightened personality did come with a hitch, though. Compared with previous Granite Mountain crews, 2013's hotshots were untested. Before taking over as superintendent, Steed had been captain for only one fire season. On other hotshot crews, it can take a captain a decade to make the jump. Filling in for Steed as captain was Tom Cooley, who had worked on the line with fewer than half of the hotshots. Nobody questioned Cooley's experience. During his fifteen fire seasons, he'd fought hundreds of blazes in many different command capacities, but now his full-time job was as a structural firefighter with the Prescott Fire Department.

Cooley had come to the hotshots in April after learning of Marsh's shoulder injury. Without Marsh, or somebody with comparable experience, Granite Mountain lacked the minimum qualifications needed to retain their hotshot status. Cooley, whose ample wildland certifications were up to date, agreed to temporarily transfer to the hotshots to help fill the experience gap. He was the number two in command of a crew he'd hardly worked with at all.

And it wasn't just Cooley and Steed in new positions on Granite Mountain. Three of the hotshots' fire experience was limited to the Hart Fire and the uneventful prairie burn, and Bob Caldwell hadn't yet spent a year as squad boss. Hotshot crews deal with a considerable amount of turnover every season, but rarely this much. How Steed and Granite Mountain would respond to its first truly dangerous assignment remained an open question.

Steed led the hotshots at a hammering pace up the road Clayton and Farnsworth had driven a few hours earlier. The slop-over rapidly spread to nearly a hundred acres, but around midnight, the humidity approached 30 percent. The extra moisture in the air damped the fire's spread to a crawl. If the crew cut line fast enough, Granite Mountain could catch the slop-over by daybreak. Thompson Ridge could still be contained.

At the top of the hike, Steed gave a quick briefing on escape routes and safety zones. It echoed the briefings he'd given all year: If the fire intensified, Steed said, step into the cold black—the already burned fuel just inside the fire's spreading edge. But this time, Steed had something else to say, too.

Maybe his earlier refusal to immediately chase the slop-over made him want to show the operations chief that Granite Mountain had the guts and firepower to catch the spot before shift's end. Or maybe he wanted to prove to himself, the hotshots, and all the crews on Thompson Ridge that Granite Mountain was as good under Steed as it was under Marsh. Whatever it was, Steed added onto his regular briefing a few words designed to pump up the crew. This was the first time the crew had really had a chance to prove what they could do on the fire line. Steed wanted them to catch the slop-over *that night*.

Bunch and the sawyers ripped their chainsaws to life and started clearing a line right on the fire's edge.

"Back-cutting! Down the hill!" Bunch hollered as the crew dived into the woods. An enormous dead pine groaned as it leaned over and popped when it separated from its stump. When it hit the ground, every hotshot within a few hundred yards could feel the tremendous thud through the soles of their boots. Granite Mountain was off and hustling to catch the slop-over.

The saws kept screaming and the swampers followed close behind, tossing still-burning logs back into the fire and pushing armfuls of cut green brush farther from the burning edge. Throughout the shift, Steed moved ahead of the crew, devised plans, then returned and worked right alongside his men.

As lead Pulaski, Donut led the way for the ten other hotshots in the scrape. His first tool strokes were nearly on the gravel road where the slop-over had jumped the lines. Starting fire lines in areas of nonflammable materials—water, rocks, roads, the black—is known as anchoring or creating an anchor point. The tactic is used on all lines to ensure that the flames can't flank or get behind the firefighters.

Within minutes, Donut moved into the pine needles and aspen leaves smoldering on the forest floor. The scrape's line traced the slop-over's creeping perimeter. Flames crawled up the sides of long-dead and still-standing trees—snags—and the woods flickered around them. Donut put the line so close to the flames that he could feel the heat on his face.

For hours, the scrape swung their tools. Every so often Donut broke his rhythm—*swing, step, swing, swing, step*—and turned around to watch the men's headlamps bobbing behind him. At first, the rookies in the scrape cut line with the mad energy of inexperience, pouring into each tool stroke every calorie they had in their bodies. But after a few hours had passed, Donut glanced back and saw that most of the headlamps were shining down at the men's boots. Fatigue was setting in. It wasn't a good sign. Daybreak was still hours away.

"Hey, Donut," Woyjeck said. "Can I take a piss?"

Donut sighed. After Chris left, Donut filled in as lead firefighter, the second-in-command on Alpha. The position allowed him to sit in the front of the buggies—a nice perk. But it also required fielding Woyjeck's many questions.

"Dude, I don't care if you piss, drink water, or sit and eat a snack," Donut said. "Just do your fuckin' job and don't go tits-up on me."

If Marsh were here, Donut thought, *he'd be bawling at the rookies to pick up the pace.* They already looked spent. The rumor among the hotshots was that Marsh was taking over the crew again as soon as they returned from Thompson Ridge. His shoulder had healed. Donut considered it good news. He felt that Steed's mellower style of leadership came at the cost of the perfection that Marsh demanded of Granite Mountain. Maybe when Marsh returned, the new guys, as Chris MacKenzie would often say, would learn.

Brendan "Donut" McDonough had learned four years earlier. His first hard firefight came just a few weeks into his rookie season. A helicopter had dropped Granite Mountain on an Arizona ridgetop within sight of Mexico. One California hotshot crew had been camped alone in the meadow for ten days. They'd eaten nothing but MREs, and there was nothing to do but work from dawn until well past dusk. The camp was like a scene from *Lord of the Flies*: Tents cluttered the meadow, empty boxes of MREs were stuffed with trash, and the other hotshots were filthy, exhausted, and fast growing irritated with one another. With no resupply, the men were forced to ration their tobacco. Before Granite Mountain flew in, a few of the veterans had the good sense to stash extra logs of chewing tobacco in the sling loads— gifts for the other crew. But not even tobacco could alleviate the suffering.

Temperatures topped a hundred degrees every day. The fire had ripped through brush that was twice the men's height but not thick enough to provide any meaningful shade. For ten days, Granite Mountain worked sixteen-hour shifts in this sweltering tunnel, clearing a ridgeline of fuel in preparation for a burnout that they ultimately lost. This job was as hard as hotshotting got, and harder still for Donut.

Donut had moved to Prescott from San Diego, California, when he was thirteen. His new school was dominated by cowboy kids dreaming of one day riding in Prescott's Frontier Days, the world's oldest rodeo. Donut stood out. He was a lanky blond skateboarder with a fondness for brightly colored tank tops and an inability to ignore insults. Donut and Bob Caldwell, who grew up in Prescott, had nearly fought twice in school; once after one of them supposedly threw a rock at the other's truck, and again when Bob accused Donut of cheating in a cooking class.

To hear him tell it, Donut faced the same exclusion when he started on Granite Mountain. He was a punching bag. Marsh hired him less because he was a stellar applicant and more because he needed to fill a position quickly after another guy had quit. Unaware that Marsh was even looking to hire, Donut had gone to the station in search of work. Marsh interviewed him on the spot, asking him his typical suite of questions. Donut earned his job because of his response to this one: "Why do you want this job?"

He was studying to be an EMT at the time, but sleeping off hangovers during most of the classes. He had been on probation for theft, had a history of smoking weed and drinking, and, most important, had a baby daughter. Donut told all of this to Marsh.

"She's why I came here," Donut told Marsh. Being a dad made him want to straighten up. "If you hire me, I'm not going to quit."

Marsh believed in giving people second chances. He hired Donut on the spot.

For most of Donut's first fire season, Bob, along with most of the other hotshots, ignored him, but when somebody did say something to Donut, it was usually scathing. Donut earned his nickname because nobody bothered to address him as Brendan or McDonough. He was McDonald's, McDoNothing, McDipshit. "Donut" stuck. Though the other hotshots seemed intent on making Donut feel unwanted and unliked, he refused to quit.

The fire on the Mexican border proved to the rest of the crew that he deserved to be a hotshot. From steep hikes and long days of swinging Pulaski, Donut bled where his pack straps cut into his shoulder and his hands, even through the gloves, bubbled with blisters. Then he caught the flu. At night, he'd lie on his mat shaking with fever and exhaustion, coughing until he vomited. Finally Marsh, who'd already had two guys flown off the line because of the bug, told Donut it was time to go to the doctor.

"I'm not leaving, Eric," Donut said. He needed the money for his daughter.

"If you're going to be a dipshit about this, you're sleeping in the corner of the meadow," Marsh said. "Because your coughing is keeping everybody else up."

Through the sickness and the hazing, Chris MacKenzie berated Donut more than anybody. "So you want to be a hotshot, huh?" was among the nicest things he said to Donut when he dropped off the back of the line during hikes. Chris's hazing got so bad that they nearly came to blows. Donut finally had to ask his squad boss to make Chris back off. After that, Chris came up and talked to Donut.

"It's nothing personal," Chris said to Donut, by way of apology. "I got it bad when I was getting raised up as a rookie. This is how you learn."

"Dude, why, though?" Donut asked. "Making me feel like a piece of shit doesn't make me want to hike any faster."

Bizarrely, the confrontation made them fast friends. Chris taught Donut little tricks to make the job easier: carrying a Clif Bar in his breast pocket, stowing aspirins in his line gear, clipping a carabiner to the outside of his pack to make setting up a tarp quicker, packing Tylenol PM on every tour. The little bit of acceptance made the job's punishment more palatable, and the two hotshots ended up moving in together after Donut's first season. Since that heinous fire on the Mexican border, Donut had operated under the belief that the fire line was the only place a hotshot could prove his mettle.

As far as Donut was concerned, the slop-over on Thompson Ridge was the first time 2013's rookies had been truly tested on the fire line. Dawn's gray light warmed the eastern slopes of the Jemez Mountains, and the line spooled out behind Donut and the scrape. Each tool stroke sent a shudder up Donut's long arms. He was tired. Everybody was. Donut paused between strokes to switch off his headlamp and drink water.

The rest of the scrape kept cutting up the New Mexico dirt. Grant, who seemed to be growing into the job's physical demands, was chugging along just fine. Sean Misner, another rookie on Bravo, was wearing thin and Woyjeck was so exhausted that at the end of the shift, two other hotshots would have to carry him off the line.

Donut had always been content to let others lead. But with Chris out because of his mom's brain tumor, Clayton working as a task-force leader trainee, and Steed scouting ahead, Donut felt duty-bound to step up.

"Let's finish this thing off! Swing those fuckin' tools!" Donut hollered from the front of the scrape, his voice tinged with a shrillness that gave him an edge of insanity. He bent harder into his Pulaski strokes, the dirt flinging behind him as if blasted by a leaf blower, and picked up the pace. "Welcome to hotshotting, boys!" he said.

CHAPTER 13

STEED'S CREW

Renan was still in a wheelchair when a blast of heat blew into the air-conditioned lobby of an Albuquerque hospital and Eric Marsh walked through the sliding doors in his Nicks-brand fire boots and tan Nomex, his handlebar mustache thicker and prouder from a month's growth. Renan hadn't known Marsh was coming to pick him up to drive him back home to Prescott, but he wasn't thrilled to find out.

Dammit, Renan thought. *I came to Granite Mountain to win the Rookie of the Year award, and now the man who hands them out is bringing me home from the hospital.*

He felt ashamed. Plus it was a long drive back to Prescott, and Marsh wasn't exactly a conversationalist. He was the guy that Renan did everything in his limited power not to upset.

"I'm glad you're all right," Marsh said, patting him on the leg. "You gave us quite a scare."

The hospital tests revealed something different from what Renan had suspected. It wasn't a return to the illness that had haunted him in his youth. Pure exhaustion had triggered intense and painful muscle spasms that caused him to lose consciousness—but it was only muscle spasms. It would take time for the buildup of lactic acid to leave his gripped muscles, but, being young and fit, he was otherwise healthy. Marsh loaded Renan into the truck. As they hummed westward through the open desert, the smoke column at Thompson Ridge grew smaller in the rearview mirror.

They talked about the potential of Renan's returning to the hotshots. It wasn't possible that summer. Given his past, Renan presented too great a liability for the City of Prescott, but depending on how his recovery went over the next few weeks, Marsh offered him a chance to work on the chipping crew—the three or four men who spend their summer creating defensible space around Prescott.

"I'd like some time to think about it," Renan said, a little disappointed.

Eventually the conversation drifted away from fire, and Marsh opened up. He told Renan about the "charity-case horses" he had picked up on the Navajo Reservation. One, a paint called Honey, had a diabetic foot and couldn't be ridden, but nor could Marsh stand to get rid of him. He loved the horse too much. Renan had never seen Marsh so relaxed. He put Renan at ease, making him laugh with sly and slightly off-color jokes and a friendly candor that seemed contradictory to the superintendent the veterans describe in stories.

This was the Eric that his friends and family knew. Life had improved for Marsh after Granite Mountain earned its hotshot status. The sting of his second divorce eased, and he fell in love again. Marsh met his third wife, Amanda, at a Denny's in 2009 during a date with a different woman, who had asked her friend Amanda to join her as the third wheel just in case things got awkward. The date went well—just not as planned. It was Amanda and Marsh who hit it off. Amanda shoed horses for a living, and with her plainspoken language and country ways, she rekindled Marsh's love of horses, a passion that had lain dormant since his youth.

A week later, they were dating regularly, and within a year they were living together at her horse ranch on the outskirts of town. Marsh proposed to Amanda during an ice-climbing trip to the fourteen-thousand-foot peaks of southern Colorado and, since then, their relationship thrived in winter—it was the season when they could spend time together.

"Eric was ninety percent hotshot and ten percent mine," Amanda would say, but she learned Marsh's dedication to his job the hard way. When they were first married, she would shift around appointments with clients to spend time with Marsh when he returned—at random and for only a couple of days—from fire assignments. But trying to cram six months of marriage into two days every few weeks was intense, unrealistic, and exhausting. The highs were dizzying, but so were the lows. The time away brewed insecurity and unattainable expectations about how their relationship should have functioned during the limited time they shared together.

After the first summer of their marriage, Amanda accepted the reality that winter was their season together. The next fire season, when Marsh returned from assignments, Amanda would keep working and let her husband relax as he preferred: with a pint of Ben & Jerry's ice cream and *Family Guy* on the TV.

Marsh's job performance improved dramatically after he met Amanda. Reviews that once straddled the line between "does not meet expectations" and "meets expectations" became almost exclusively "exceeds expectations." One firefighter he worked with even saw fit to write a letter to Marsh's supervisor lauding his performance: "He motivated without belittling. He had a good humor and knew when and how to use it. He believes in the 10 and 18 and understands those guidelines were born out of the ultimate sacrifice."

But what Marsh hadn't anticipated was that the greater wildland community still didn't welcome Granite Mountain into its ranks, even though it was a sanctioned hotshot crew. The other hotshot superintendents in Arizona and New Mexico treated Marsh as the outsider. That resistance was commonplace for the culture.

"One superintendent I worked with didn't talk to me for three years," said Jim Cook, the former superintendent of the National Park Service's Arrowhead Hotshots, one of the first non–Forest

Service/BLM crews to join the hotshots' ranks. "The first time he spoke to me was after Arrowhead spent all day putting in this heinous piece of line. The supe walked past my crew and the only thing he said to me was 'Nice line.' It was a huge breakthrough."

If Marsh was treated with isolation, so was Granite Mountain. He responded to other crews' quiet belittling of his men by reaffirming his pride in their job: If his men believed in their work, they'd do it well. He regularly told his crew, "You're the best hotshots in the nation."

While many longtime firefighters felt that Granite Mountain was a good hotshot crew, they still believed that the best crews could be run only by the most experienced superintendents. Time on the line is synonymous with a firefighter's ability to anticipate a fire's next move and direct his or her crew accordingly. Some superintendents have almost forty years' experience fighting wildfires, with more than twenty-five at their crew's helm. By comparison, Marsh had a moderate level of experience. He'd been a superintendent for a total of nine years. Granite Mountain was a hotshot crew for five of those.

During that time, Marsh had developed a reputation for inconsistency on the line. He refused assignments more senior superintendents readily accepted, while taking on aggressive assignments his seniors turned down. On Arizona's Horseshoe Two Fire, in 2011, Granite Mountain's third year as a hotshot crew, Marsh agreed to back-fire off a densely vegetated ridgeline, an assignment that more senior superintendents on scene had refused. Granite Mountain and the other crew they were working with lost the burnout. All crews eventually lose burnouts—it's one of the job's unavoidable realities— but other hotshot superintendents on the Horseshoe Two Fire commented that Marsh's decision-making on the line had been suspect.

Later, on a blaze in Idaho, Marsh refused an assignment that he felt would have exposed the crew to many trees that were likely to fall with a strong gust of wind. In the aftermath, the division filed an evaluation stating that Granite Mountain had failed to meet the objectives. Marsh's response to the negative review was defensive and exceptional: He handwrote a six-page letter to the incident management team detailing the circumstances from his perspective. His point, which he made with ample exclamation marks, was clear: "Resources need the ability to refuse unsafe assignments without fear of reprisals. This is not about me vs. Div A, but rather about a crew being punished with a bad evaluation for trying to go home safe!"

Some of Granite Mountain's hotshots took note of how seriously Marsh seemed to take himself and his profession. One time, when he'd shown up to the station wearing a bandanna around his neck, one of the squad bosses complimented Marsh on his "cute scarf." He recoiled and corrected him.

"It's a neckerchief," Marsh said.

Privately, this only made the guys laugh harder. A few of the more senior firefighters, Steed among them, poked fun at Marsh behind his back.

By 2011, Marsh was looking to move beyond the superintendent position. He was forty-one and felt he was getting too old to handle the constant pounding of the fire line. Marsh set his sights on Wildland Division chief, a position that would put him in charge of planning defensible space for Prescott. With the retirement of that division's current chief, Darrell Willis, on the horizon, the fire department went as far as hiring Granite Mountain a new captain, a Forest Service veteran hotshot named Aaron Lawson, to allow Marsh to eventually transition into the chief's job. Lawson was well qualified to take over as Granite Mountain's superintendent.

While the plan sounded good on paper, Marsh and Lawson never got along. Willis chalked it up to the fact that Marsh was "one of the most ethical and loyal people" he knew. While it isn't clear if Willis's observation was about what an upstanding person Marsh was, or a subtle jab at Lawson, one thing was evident: Lawson was at a disadvantage from the start.

After years of simmering tension and one-upmanship among Granite Mountain, Forest Service, and BLM hotshot crews, Marsh had reached the conclusion that the Forest Service was dominated by ego-driven, petty, and aggressive firefighters who tended to act brashly for glory and viewed any assignment other than directly fighting fire as beneath them. In a document describing what set Granite Mountain apart from federal crews, Marsh wrote, "In a workforce dominated by Forest Service and other federal hotshot crews ... we are odd. We look different. Not because our buggies are white instead of green, but because we smile a lot. We act different. We are positive people. . . . Our folks are smart, motivated, and highly trained professionals that don't see any task as 'beneath them.'"

At the time of Lawson's hire, Willis was pushing Marsh and the crew to adopt what he called "the Prescott Way," his overarching philosophy that the hotshots' first priority was to always act like the civil servants they were. During the winter, Granite Mountain's overhead plowed residents' driveways and hung Christmas lights around the courthouse square. In summer, the whole crew did fuels treatments around homes. Lawson certainly wasn't against helping people when they needed it—he'd long done that during his tenure on Forest Service crews—but the sheer volume of Granite Mountain's charitable and community-oriented tasks must have come as a surprise to him.

Whether it was Lawson's inability or unwillingness to adapt to the Prescott Way or simply a personal issue, the friction between the two men only continued to mount, and Marsh was clearly relieved when Lawson quit after just two seasons. In his June 2012 self-appraisal of his job performance, Marsh wrote that he was proud "to keep the crew focused while undergoing a personnel issue with my assistant. It was difficult to not be angry and vengeful in that situation." Until Marsh could find a suitable replacement, he'd remain Granite Mountain's superintendent.

Granite Mountain hadn't caught the slop-over on Thompson Ridge. By the time the sun rose, it was apparent that they'd fallen just short of finishing the job. But in the morning, a Forest Service hotshot crew took care of the rest in a few easy hours.

The operations chief who designed the burnout later apologized to Steed for pressuring him into chasing the fire. The subdivision of homes was no longer threatened. Nor was Los Alamos. And the local press, which had covered the burnout, knew nothing of the few thousand extra flaming Ping-Pong balls that had supposedly been launched into the forest. Despite the burnouts, the vast majority of the backfire had actually turned out to be the equivalent of a prescribed burn pulled off during the peak of fire season. Some of it had burned hot, but most was low-intensity. Thanks in part to Steed and Granite Mountain, Day's management team could proudly point to its accomplishments on Thompson Ridge. Over the next few decades, wildfires in the area should pose less of a threat.

Day's management team started releasing some of the six hundred firefighters on scene, but Granite Mountain wasn't going home yet. They were kept on with a few other crews to monitor the fire. For the men, that meant mopping up, punctuated by a few shifts burning out pockets of green fuel still inside the line. Like a well-earned sigh after a series of testing and exhausting shifts, the downtime during Granite Mountain's last days on Thompson Ridge provided opportunity for great fun and small dramas.

Before heading out for one night shift, Steed brought all the hotshots to an After Action Review, a group discussion among the overhead of the engines, crews, and helicopters involved in the burnout. Granite Mountain was the only full crew there. The operations chief and division walked these forty or so firefighters through the original plan, then the superintendent or captain of each resource involved debriefed about his or her crew's experience. Steed intended for the After Action Review to serve as a learning opportunity for the hotshots, which perhaps it did, but most of the men saw the meeting as the setup for a long string of locker-room jokes.

Later that night, the hotshots assigned themselves roles and held an impromptu skit mocking the After Action Review. They drew a penis-shaped map of the fire and hung the sketch on the side of a buggy. Grant played the operation chief, the star. He opened the

Granite Mountain Hotshot superintendent Eric Marsh (red helmet) filling gas canisters during a burnout operation. Marsh's role in the fatalities at Yarnell Hill was highly controversial. Some longtime firefighters counted him among the best firefighters they knew; others called him a "bad-decisions, good-outcome guy." Jakob Schiller

Clayton Whitted, a former youth pastor, helped shape Granite Mountain's uniquely religious character. Whitted was among the crew's supervisors on June 30, 2013. Jakob Schiller

Hotshot Brandon Bunch, twenty-two, faced a tough decision in late June: stay on with the crew to make much-needed money for his family, or return home to be with his wife, Janae, for the birth of their third child.   JAKOB SCHILLER

Bob Caldwell's intelligence was said to be high enough to merit membership in Mensa. In the minutes before the fire claimed nineteen firefighters' lives, he was one of three hotshots who communicated with the Air Attack plane. The radio transmissions pre-ceding Caldwell's became known as "the mayday call." Jakob Schiller

Afghanistan war veteran and lead firefighter of Bravo squad Travis Turbyfill sets a backfire in New Mexico. Jakob Schiller

The column of 2011's monumentally destructive Las Conchas Fire darkens the skies above the town of Los Alamos, New Mexico. For fourteen hours straight, the megafire torched an acre of pines every 1.17 seconds. KRISTEN HONIG/VALLES CALDERA NATIONAL PRESERVE

A cabin threatened by the Thompson Ridge Fire in the early hours of June 5, 2013. Three weeks before Yarnell Hill, the Granite Mountain Hotshots helped save this cabin, and others like it, from burning in New Mexico's Valles Caldera National Preserve. KRISTEN HONIG/VALLES CALDERA NATIONAL PRESERVE

Fully half of the nation's air tanker fleet was fighting the Yarnell Hill Fire on June 30. Here, a specially converted DC-10 supertanker unloads eight thousand gallons of retardant. In the foreground are Granite Mountain's buggies; the knoll in the background is Brendan "Donut" McDonough's lookout.

For reasons that remain unclear, Granite Mountain left the safety of the already burned vegetation for a ranch near the town of Yarnell. Minutes before the fateful decision was made, hotshot Scott Norris texted this photo to his girlfriend, Heather Kennedy, along with the following note: "This fire is going to shit burning all over and expected +40 hr wind gusts from a t-storm outflow. Possibly going to burn some ranches and house." Scott Norris

This photo of the smoke column was shot around 4:40 P.M., within minutes of superintendent Eric Marsh's last radio transmission. At this point in the day, fifty-mile-per-hour winds were stoking the Yarnell Hill Fire and the blaze had grown by approximately 2,700 acres in just forty minutes.

Courtesy of the National Interagency Fire Center

The view from an Air Attack plane circling the Yarnell Hill Fire shortly after the hotshots were burned over. The white smudge in the upper left corner of the picture is the Helms' place. The small clearing just to the left of the wingtip is where Brendan "Donut" McDonough considered deploying his fire shelter, an aluminum blanket designed to deflect heat but not withstand direct contact with flames.

Courtesy of the National Interagency Fire Center

As the Granite Mountain Hotshots deployed their shelters, the fire slammed into Yarnell, burning more than a hundred homes. When the flames hit, many of the town's 650 residents had not yet evacuated. Shot by the Air Attack.

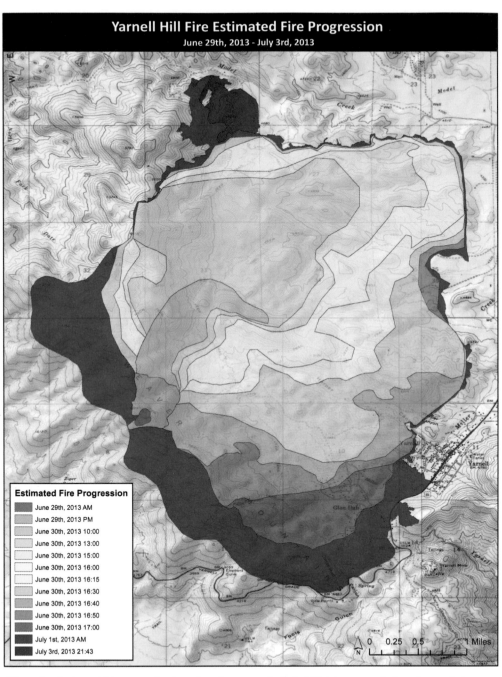

Yarnell Hill Fire Estimated Fire Progression
June 29th, 2013 - July 3rd, 2013

Estimated Fire Progression

- June 29th, 2013 AM
- June 29th, 2013 PM
- June 30th, 2013 10:00
- June 30th, 2013 13:00
- June 30th, 2013 15:00
- June 30th, 2013 16:00
- June 30th, 2013 16:15
- June 30th, 2013 16:30
- June 30th, 2013 16:40
- June 30th, 2013 16:50
- June 30th, 2013 17:00
- July 1st, 2013 AM
- July 3rd, 2013 21:43

0 0.25 0.5   1 Miles

Note the fire's rapid increase in size after the second wind shift hits around 16:30 (4:30 P.M.).

ARIZONA STATE FORESTRY

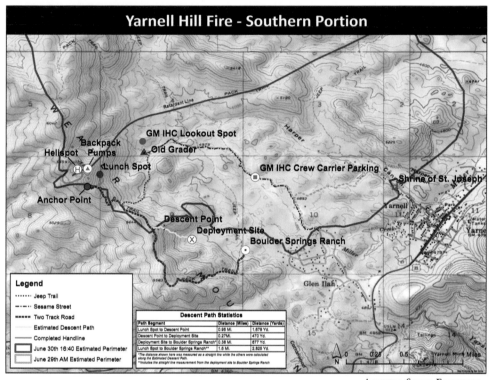

Yarnell Hill Fire - Southern Portion

Legend

· · · · · · · Jeep Trail

— · — · Sesame Street

▬ ▬ ▬ Two Track Road

——— Estimated Descent Path

——— Completed Handline

☐ June 30th 16:40 Estimated Perimeter

☐ June 29th AM Estimated Perimeter

Descent Path Statistics

Path Segment	Distance (Miles)	Distance (Yards)
Lunch Spot to Descent Point	0.95 Mi.	1,678 Yd.
Descent Point to Deployment Site	0.27 Mi.	470 Yd.
Deployment Site to Boulder Springs Ranch*	0.38 Mi.	677 Yd.
Lunch Spot to Boulder Springs Ranch**	1.6 Mi.	2,825 Yd.

*The distance shown here was measured as a straight line while the others were calculated along the Estimated Descent Path.
**Includes the straight line measurement from the deployment site to Boulder Springs Ranch

ARIZONA STATE FORESTRY

Investigators at the site where the hotshots deployed their fire shelters. It was roughly the size of a tennis court. The Helms' place, the hotshots' intended safety zone, is the unburned space on the center-right side of the picture. The burned debris beside the rock pile in the foreground is the hotshots' equipment.   WILLIAM FOLEY

The blackened contents of one hotshot's pack and rhino tool. Note that the wooden handles of both tools are completely burned and the pack's cloth is nearly gone. Experts estimate that flames taller than sixty feet swept through the basin, and temperatures exceeded two thousand degrees.   WILLIAM FOLEY

The Helms' place, Granite Mountain's intended safety zone, after the flames passed through. Though the Yarnell Hill Fire's intense heat cracked the buildings' windows, the ranch survived because the homeowners had cleared their property of brush before the fire started. JOHN WACHTER

The view from the two-track road the hotshots used as an escape route to their safety zone, the large ranch in the background. Near the point where this photo was taken, the hotshots left the two-track road and moved cross-country toward the ranch, which was .6 miles away. Hampered by thick vegetation and steep terrain, Granite Mountain fell roughly a third of a mile short of the safety zone. JOHN WACHTER

An aerial view of the Helms' place and the basin where the men died. A bulldozer built a road into the deployment site late on the night of June 30.   ARIZONA STATE FORESTRY

Looking up at the basin from the Helms' place. The hotshots' descent route is the small, steep draw that ends at the flagpole. Note the ridge on the right side of the image. These massive granite boulders blocked the men's view of the fire.   JOHN WACHTER

Nineteen white hearses transported the men from Phoenix, where they were autopsied, back to the hotshots' home base in Prescott. In temperatures that exceeded a hundred degrees, thousands of people lined the road to offer their respects to the fallen firefighters. LAURA SEGALL/GETTY IMAGES

Brendan "Donut" McDonough, Granite Mountain's sole survivor of the Yarnell Hill Fire, addresses a crowd of 14,400 people on July 9, 2013, at a memorial service held outside the crew's home base in Prescott. In the foreground are alumni of the Granite Mountain Hotshots; Vice President Joe Biden looks on.   DAVID KADLUBOWSKI—POOL/GETTY IMAGES "meeting" with a mopey confession: "First, I take full responsibility for this. This is my fault. This is all my fault." Then the others stepped in.

Scott assumed the role of another hotshot superintendent on scene who happened to be Native American. He adopted the stereotypical accent and joked about the crew having to use "smoke signals to coordinate with Granite Mountain" because "the flame lengths were taller than our tepees." The only thing that made the men laugh harder was Donut's impersonation of Steed. He sped up his voice to hyperactivity and talked about when the "Gatorades got burned up," then paused mid-discussion to knock out a few push-ups "to tone my bis and tris" and improve his "beach bod." The hotshots howled.

After the high of the show, the men headed back onto the line and spent the night mopping up hot spots. Bravo squad's rookies Sean Misner, Woyjeck, and the bearded redhead, Dustin DeFord, were continuing their endless conversation about things they couldn't have—showers, steaks, sex, beer—when Bunch walked up and asked, "Who the fuck took my chips?"

He'd been saving them as a late-tour delicacy ever since Grant bought him the chips and dip from the gas station more than a week earlier. Somebody had taken them from his basket in the buggy. The rookies were silent for a moment, then Misner and DeFord both blurted out, "Woy-shack!" Everybody on the crew deliberately mispronounced Woyjeck's name.

Bunch pivoted.

"Dude, I'm so sorry. I really didn't know," Woyjeck said. "Steed told me to get him some chips from the buggy, and I saw that bag in the front seat and thought those were the ones he was talking about"—in fact, they probably were; Steed wasn't above a little espionage—"so I grabbed them."

"Those were my fucking chips," Bunch said, stifling a smile at, considering the crime, Woyjeck's incommensurate fear.

"I owe you. I'll get a bigger bag—a couple; no, a few!—when we hit the next gas station," Woyjeck said. "I promise. Dude, I'm so sorry. I know you love chips."

Bunch couldn't help but enjoy watching Woyjeck grovel. The sight was almost worth losing the chips for. But poor Woyjeck felt taken advantage of. He was a polite kid with a nervous constitution. His dad was a structural firefighter, but the ego and bluster of young hotshots was still unfamiliar to Woyjeck. The twenty-one-year-old had grown up in Orange County, California, and dreamed of following in his dad's footsteps since he was a kid. By age twelve he had his CPR certification, by fifteen he was an EMT, and he'd spent one year fighting fires on a Type 2 Initial Attack crew in South Dakota. As with Grant, the hotshots were a step toward his goal of becoming a structural firefighter. But the time away wasn't coming easy for him, either. He called his mom and dad almost every day from the fire line.

Woyjeck didn't mind jockeying for position and acceptance. He liked and excelled at the physical tests. But the constant posturing was tiresome and against his nature. Later that night, Woyjeck and Grant worked together on a quiet piece of line.

Grant knew that Woyjeck's lease was running out and that he needed a place to live in Prescott. "We've got a spot opening at our house," he told Woyjeck. "Take it."

What Grant didn't tell him was that the roommate had left because Grant had constantly badgered him about not meeting Grant's strict cleanliness standards. In any case, Woyjeck was thrilled. Living with Grant was an easy fix to the housing problem, and it seemed to be a step toward solidifying their friendship. But later that night, when the crew staged a wrestling match, Grant took the opportunity to assert his dominance—another act of posturing. Somehow Woyjeck, the slightest guy on the crew, had handled a few of the older and bigger veterans. He was feeling good when he came to his future roommate. With the other hotshots forming a circle around the wrestlers, Grant quickly undercut Woyjeck's legs and pinned him with skill and savage enthusiasm. Then Grant made the point he'd been trying to make in his confrontation with Tony a week earlier on the Hart Fire: He might be a rookie, but he wouldn't tolerate disrespect.

"Now you know who the man of the house is," Grant said, loud enough that the hotshots surrounding the wrestlers whooped collectively, as if to add an exclamation mark.

Bunch sat alone watching Thompson Ridge burn itself out. He'd sharpened and cleaned his chainsaw and the chains. After that, there wasn't much to do but think about Janae, Ben, Jacob, and the baby who was coming soon. Bunch had three more weeks on the fire line before he left the crew in the beginning of July.

Bunch hadn't talked to Janae since June 11. The crew had been heading back to their sleeping area, and the short ride in the buggy provided the only time the hotshots had to talk to their families—the campsite didn't have service. The back of the buggy was a clamor of unconnected conversations. There always seemed to be some tidbit of news that came out of their brief calls. One day Scott learned that his sister had given birth.

"Congratulations Mama!!" Scott texted. "I'll call you after we eat in a little bit. So proud of you and love you bunches." His sister had never known Scott to be so emotional. Another day, the crew heard from Chris. His mom's surgery had gone well. She was in remission, and Chris planned to rejoin the crew when the hotshots returned to Prescott.

That day, Bunch's conversation with Janae didn't include any news. She just told him how things were going. She never actually used the words "Please come home," but it was the only thing she was really saying to him. He hadn't realized it at the time, but Bunch heard it now.

The pregnancy plus two boys was getting to be too much. Jacob had been throwing fits without his dad around, and Ben, still in diapers, demanded more attention than Janae could provide by herself. She had taken the kids to her mom's house and was staying there, but that added a new and different dimension to her stress.

Bunch's time away from the family was nothing new to Janae. They'd gone to high school together—Bunch was a wallflower; she was the wild one—but hadn't started dating until after they'd graduated and bumped into each other at a local coffee shop. Janae had given birth to Jacob months earlier and was now a single mom. She was reading the Bible—a rare quiet escape from the madness of the past few months—when she saw Bunch walk in. Small talk wasn't Janae's thing, or, for that matter, Bunch's. She turned her back to avoid eye contact, but Janae "kind of had to look up when Bunch stopped directly in front of me."

Weeks later, he was telling her about his love for hotshotting. She thought the flames sounded impressive and the job dangerous, maybe even a little sexy, but she wouldn't realize how uncomfortable dating a hotshot could be until the summer of 2011. That year, Bunch was gone on assignments 120 days. He never tired of sleeping under the stars. When the crew finally got days off, Bunch would set up camp in the backyard, and the family would fall asleep under Prescott's clear night sky.

On Thompson Ridge, Janae called her troubles with the kids "normal fire-line whining," but both of them knew that when Bunch left the crew in a few weeks, he wouldn't have a job. With the new baby coming soon and two kids to deal with already, their savings wouldn't last. A few more weeks of work could mean a few more grand.

Janae and the boys could wait, Bunch told himself. They'd have to.

Donut was holding line with Clayton and Wade. It was nearing 3 A.M. and, as sometimes happened, Clayton started sharing lessons from the Bible.

"I don't want to be a Christian until I'm ready to be a *good* Christian," Donut said. He toyed with his Pulaski in the white ash of a fire long since cooled. "And honestly, I don't think I'm ready to be a good Christian."

"You know you can still make mistakes," Clayton said. "It's not like you sign up and have to live a perfect life forever after or you end up in hell. It's just that after accepting God into your life, you find yourself *wanting* to live better."

This made sense to Donut. In many ways, a relationship with God sounded similar to his relationship with Granite Mountain. After coming to the hotshots, Donut had made a concerted effort to refocus his life around habits that were far more productive. It wasn't the easiest of transitions. Once, while the men were returning home from fighting fires in Idaho, they stopped in Las Vegas for the night. Marsh rented rooms for the men near the Strip. The hotels were cheap, but putting Granite Mountain up close to the action was Marsh's subtle way of rewarding the crew with a chance to let loose. Before letting the hotshots go, he asked the men to be back at the buggies at 7 A.M., an hour later than normal.

Donut was later still. He wasn't twenty-one yet, but the older guys bought him booze and then happily disappeared to the bars, leaving Donut to find his own adventures. Sometime after sunrise, he stumbled back into the hotel with a collection of new stories—true, false, or exaggerated—that involved a stripper's lap-dance offer to sleep with him, a fall from a moving train, and a men's-room spat with a man in a wheelchair. The night culminated back at the hotel, where, Donut said, a lesbian couple licked him simultaneously. When he got back to the buggies, he had some of the best stories the hotshots had heard in years.

Donut had never taken part in church traditions, but if what Clayton said was true and God really could forgive his behavior, Donut could consider Christianity.

"God forgives," Clayton reassured him. Then Clayton and Wade took Donut's hand, and in the soft glow of the dying fire they prayed for Donut to come to God.

Given Renan's incident, the lack of sleep, and the urgent slop-over, Grant's first full two-week fire tour had been harder than most hotshots', which was probably why he felt so poignantly disappointed when Steed told the crew that they had an opportunity to spend another two weeks on the line without going home. A new fire in southern New Mexico was growing rapidly, threatening to become the largest fire in state history. The incident commander needed crews.

The hotshots were gathered behind the trucks at fire camp, leaning on bumpers or squatting while staining the ground with tobacco spit. Steed laid out the options. Because they'd already worked two weeks straight and the SWCC wasn't requiring them to head south, he let the men decide whether they wanted to keep rolling to fires or return to Prescott. Another fire assignment could earn the men a few thousand dollars. It also meant as many as fourteen more days away from their families. The decision was unanimous: The men wanted to go home.

Steed and the squad bosses headed into the camp to deal with the trail of paperwork—hours worked, medical incidents, vehicles repaired—that makes it possible for the NIFC to track the location, on almost an hourly basis, of the nation's firefighting resources. Filling out all of the paperwork took Steed and the squad bosses the better part of an hour. When they finished, Granite Mountain was officially released from Thompson Ridge.

The first thing Steed and Clayton did was walk across New Mexico Highway 4 to a row of trailers parked across from the fire camp. Every one of the half-dozen trailers was occupied by a vendor printing and selling T-shirts that portrayed cartoonish scenes of Thompson Ridge. Among other things, the designs included an air tanker flying low over oversize flames and hotshots working in the foreground. THOMPSON RIDGE FIRE, SANTA FE NATIONAL FOREST, MAY–JUNE 2013 was written across the top.

These vendors chase smoke columns around the West. Usually, locals or firefighters who see fewer flames than hotshots buy the shirts—enough for these vendors to make a modest living. It's taboo for hotshots to buy the shirts. The tees are garish. But Steed and Clayton had a plan; they picked the loudest tie-dye off the rack.

When they returned to the buggies, the men again gathered around.

"Before we head home, we've got a little announcement," Steed said. "Bunch has got another boy on the way. He told us a few nights ago that he needed to spend time with his family. This is his last tour with us."

Most of the men already knew as much—there are few secrets on a crew. During the slow shifts on the line, his and Janae's conversations kept replaying in his head, until Bunch finally decided to put his family above all else. Losing a few grand over the coming weeks would sting a little bit, but missing the birth of his child would stay with him forever.

The hotshots chanted his name. "Bunch! Bunch! Bunch!"

"But before he goes, we've got a little something to thank Bunch for his years of hard work," Steed said as he presented him with the T-shirt. "You've earned this."

Before the advent of cell phones, fire assignments were a lot like a sailor shipping off to sea. When firefighters left home, their families might not hear from them again for weeks. The odd calls, when they did come in, were usually made from pay phones at gas stations in the innumerable tiny ranching and mining towns that break up the empty roads of the West. Each hotshot might get five minutes in the booth to talk to a spouse or parent, deal with a late mortgage payment, or leave a heartfelt message on an answering machine before the next firefighter in line was tapping on the Plexiglas and pointing at his or her wrist. With a time limit on gas station stops, few things caused fights between hotshots like pay-phone lines.

Cell phones reduced the tension to small squabbles about charger space in the cigarette lighters. On the drive home from Thompson Ridge, the text messages flew out of Granite Mountain's buggies by the dozens.

GRANT: This drive is going to take forever. When we get
   back we have to wash the buggies, clean up . . .
LEAH: Home at 6?

LEAH: I'm cleaning the house.

GRANT: I've been sleeping in the dirt for two weeks.

LEAH: That's exactly why I'm doing it!

But for many hours, the hotshots droned through the flat, wind-swept high desert that lies between Jemez Springs and Prescott. The crew didn't pull into the base until after seven and weren't done clean-ing the buggies and preparing the trucks for the next fire until an hour after that. Grant left the station the moment Steed released the crew. Bunch, though, lingered, taking a moment to look around the place that had been at the center of his life for the past four years. There was no doubt he was going to miss the station. Small things suddenly felt sentimental—Clayton's antique chainsaws and ever-growing piles of thrift-store kitsch, a broken hand tool with TURBY TOUCHED THIS written on it, the locker his old squad had painted Granite Mountain's logo across.

Clayton was also slow to leave. He'd been Bunch's squad boss for two of his four years, and Clayton and his wife had been the only two guests at Bunch and Janae's wedding. Before that fire season started, Bunch had mentioned to Clayton that he and Janae were eloping.

Even though it wasn't technically an invitation, Clayton took it as one. Just moments before the ceremony started, he and his wife, Kristi, ran into the courthouse. "No way I was letting you get married alone," he told his buddy.

Saying goodbye was as hard for Clayton as it was for Bunch. When he saw Bunch head out to his black Toyota pickup, he whistled to him across the parking lot and jogged up.

"Hey . . ." he called. "I'm proud of you for leaving the crew for your family." Then Clayton gave Bunch a hug. Bunch, shorter than Clayton by a head, gave him a half smile and turned his head into Clayton's shoulder to hide welling tears. The goodbye had a formality to it that made Bunch's departure uncomfortably real. He pounded Clayton's back hard several times and quickly left the station.

While Grant and his crewmates finished up their duties, Leah had rushed out to get a manicure and pedicure and was a few blocks from home when Grant pulled up behind her in his Dodge Neon. Her heart fluttered. He was leaning entirely out of the window, waving with his left hand and driving with his right. Fifteen days was the longest they'd been apart.

For the couple, Grant's two days off were bliss. They slept late and ate real food, heading to dinner at a local restaurant where Grant, now old enough to buy alcohol, splurged on microbrewed beers and ate the better part of two meals for dinner. At home, Grant nested. He built a fire pit in the backyard and spent the afternoon stacking wood in a pleasant rhythmic delirium. When he was done, he put on the Forever Lazy onesie—blue adult pajamas complete with booties, which Grant's grandma had given Leah for Christmas. Leah didn't like wearing them. But being swaddled in fleece helped Grant relax. He tried to forget the stress of the past couple of weeks and tried hard not to think about the fact that he'd be back at work on Tuesday. That night, he fell asleep to *Tommy Boy*.

The days off weren't so different for the other guys. Each man tried to squeeze two weeks of normal life into two short days. Scott met his nephew for the first time. He held the baby like a vase he was afraid to drop while his sister laughed at how uncomfortable he looked. "I'm going to teach him to shoot a gun when he's three," Scott declared.

Donut spent time with his young daughter, Michaela, a toddler now, towheaded and fast learning to abuse the power of the word "no." He had dated Michaela's mom during high school, and they were no longer on the best of terms, but she watched Michaela full-time while he was away on fires, and they shared custody of the little girl when Donut returned home.

One night, after dropping Michaela off at her mom's, Donut went out to Whiskey Row. A friend of his, a semi-professional mixed-martial-arts fighter, had just gotten out of the military. After hours of partying, hopped up on cinnamon-flavored whiskey and testosterone, the two ended up in a fight with six guys. Donut fought two of them while his friend apparently took care of the other four, and they ran back to Donut's apartment higher still on adrenaline.

One corner of the living room was dedicated solely to Michaela's toys, but the place was surprisingly clean, considering that the only people who ever lived there were two young hotshots and a baby girl. Chris was trying to sleep back at home when Donut and his friend bounded up the stairs and pounded on Chris's bedroom door.

"Dude, get up!" Donut said. "It was fucking crazy."

He swung open Chris's door and immediately began pantomiming the motions of the fight—pushing, ducking, punching—so frenetically that it took him a long moment to realize that a woman was in bed beside Chris. When he finally did, Donut slammed the door shut and buckled over in laughter.

PART THREE

ONE FOOT IN
THE BLACK

CHAPTER 14

NEW STARTS

On June 18, the hotshots were back at work and had barely finished their first run in weeks when somebody spotted white smoke curling over the Bradshaws, to the west.

"That's not a good place for a fire. There are million-dollar homes up there," Donut said to no one in particular. "If they don't catch that in an hour it's gone."

It was 11:30 A.M. and already well into the eighties, with winds gusting to nearly thirty miles per hour. The fire was burning in chaparral, a mix of five or six oak-related species that cover parts of the Southwest and Southern California. When it's dry and windy, as it was now, chaparral, with its woody trunks and light and flashy leaves, acts as both the metaphoric wick and the dynamite.

Chuck Maxwell, the forecaster in Albuquerque, was watching the factors—drought, grass growth, temperatures, the monsoon—align over the Prescott area. Two days earlier, on June 16, he'd issued a warning to all firefighters in the Southwest. The season had shaped up more or less as he'd predicted it would back in May. Monsoonal thunderstorms had already extinguished the large fires in parts of New Mexico and southern Arizona, but the north and central part of the state had just entered its terrifying prime. Prescott hadn't seen a drop of rain, or cloud cover, for that matter, since the beginning of May.

The monsoon rains were still weeks out, and until they arrived, fire danger would only get more volatile. Since the 1st of June, temperatures had climbed past the high eighties every day, and the humidity consistently hovered in the teens but at times dropped as low as 4 percent. The National Weather Service had issued a Red Flag Warning, the first of many to come in the weeks ahead, and the ponderosa pines and chaparral surrounding Prescott were so dry that the probability of ignition—a calculation that measures the chance that a spark will kindle into a blaze—was at the exceedingly rare 100 percent.

Maxwell was unequivocal about the dangers. "Firefighters should acknowledge that the fire growth and fire behavior they encounter this year may exceed anything they have experienced before," he had written. "Normal strategies and tactics may need to be adjusted to account for the drought factor."

"Fucking send us already," Donut said. He ran inside to use the bathroom, and by the time he was out, Steed was yelling for the hotshots to load up.

Once on the road, Grant texted Leah.

"We're on a fire in town. I won't have my phone."

"OK be safe. I can see the smoke."

Bunch could see it, too. He was on the way to his new job, screwing cabinets together. Janae's dad had gotten him the position, which he already found boring, and, watching the column rise, milk white and growing fast, he felt a flash of guilt. He wanted to be there with his crew.

The new start was called the Doce (pronounced "*doh*-see"), after the shooting range it had started beside. It was the same name given to a blaze outside Prescott that the Perryville handcrew had fought in the same area in late June 1990, shortly before six of them died on Arizona's Dude Fire. The new fire was burning in a contiguous chaparral-and-juniper thicket that ran from its ignition point, six miles to the north, up and over the crew's namesake Granite Mountain, and straight onto the flats near the Prescott suburb of Williamson Valley. With conditions so extreme, the Doce could close that distance in a matter of hours.

Prescott has always had a fire problem. Founded in 1864, for the gold in the Bradshaw Mountains, the town in its early years claimed the precarious distinction of being the only one in Arizona built entirely of wood. (Others usually had more adobe or brick structures.) As a result, Prescott burned—regularly. During its first thirty years of existence, three separate fires razed entire sections of the town.

The city established a fire department in 1885, with four twenty-five-man volunteer brigades—the Toughs, the Dudes, the O.K.s, and the Mechanics Hook & Ladders—and installed surface wells on each corner of the courthouse square a few years later. The wells and dedicated bucket brigades still couldn't tame the flames that unnerved the miners and cowboys every summer. Doc Holliday lived in Prescott before the infamous shootout in Tombstone, and during his time there, he often knocked back bourbon in the bars on Whiskey Row. Every one of those forty bars eventually burned, but it didn't matter much. With so much gold in the hills and so many miners eager to spend it on drink and women, Whiskey Row was quickly rebuilt after every blaze.

The burns got to be so regular that when a fire sparked outside town in 1900, the drinkers at one lavish hotel lounge took the time to remove the hand-hewn oak-and-maple bar and stash it next to the slate courthouse across the street. The brick pub, known at the time as the only "absolutely fireproof building in town," scorched from the inside out. Its bar was reinstalled in the Palace Hotel, where it's still in use today. Legend has it that the same patrons who saved the bar snatched a bottle of booze on the way out, and when the flames hit they sat by the courthouse and passed the bottle while watching their watering hole burn.

By 2013, it had been more than a decade since the last serious fire, when, in one unsettling May afternoon, more than a thousand homes were threatened and 2,500 people evacuated. Flames that dwarfed three-story houses were visible from the courthouse, and glasses of bourbon and antiques store transactions were left unfinished as patrons fled beneath smoke that blotted out the afternoon sun. Ash fell so heavily it collected against curbs like blowing snow, but this time downtown's bars stayed standing through the summer of 2002.

The reason Whiskey Row was spared was Crew 7, the predecessors of Granite Mountain. They'd spent that winter cutting a fuel break on the southern edge of town. When the flames hit the defensible space, the fire sat down just long enough for air tankers and hotshot crews to lasso the blaze with line. Back then, Crew 7 could only stand by and watch. It was their first year on the job, and they weren't yet certified to do anything but thin the dense underbrush and pines on the outskirts of Prescott. But because of their work, the blaze that the incident commander predicted would destroy the town's famous strip of bars and thousands of houses was caught at just thirteen hundred acres. Only five homes were lost.

The town owed its survival to Crew 7 and, by extension, to a little-known organization called the Prescott Area Wildland Urban Interface Commission. The twenty-to-thirty-member group was a consortium of the Forest Service, the BLM, and state, county, and city firefighting agencies. For the past two decades, "PAWUIC" had been working behind the scenes to prepare for future fires. Among its founding members was Darrell Willis, who would go on to become the department's Wildland Division chief—the Granite Mountain Hotshots' boss.

As is usually the case with natural disasters, it took a near catastrophe to finally spur Willis and PAWUIC's efforts to make the city defensible. The fire that changed the city's fuels-management direction started in 1980, and though it didn't burn the town, it helped Willis recognize that with every passing fire season, the stakes increased. Droughts were becoming more common, and Arizona's population was exploding. Between 1970 and 1980, Yavapai County, of which Prescott is the seat, nearly doubled in size, from thirty-six thousand to seventy thousand. Statewide, eighty thousand new homes sprang up in rural areas. There were no codes requiring that homes be made firesafe, and most were built with wood siding and shake roofs, all too often in brush fields or beneath ponderosa pines that gave homeowners the feeling that they owned a small piece of wilderness. In one very important way, they did: The new homes were more fuel for the inevitable wildfires.

Nationwide, seventy thousand communities—some 140 million people and 40 million homes—sit in the path of fires. Though state and rural fire agencies contribute immensely, the burden of protecting these towns has fallen largely on the federal government. Once responsible for managing lands for timber, recreation, wildlife, and watersheds, the Forest Service and the BLM, the two biggest players in the wildfire business, have now effectively become federal fire agencies that watch over 180 million acres of once-rural land that development is steadily encroaching upon.

"Homeowners expect fire protection, and the government—the Forest Service, the BLM—is the only firefighting agency that's big enough to stand between the flames and their houses," said Harry Croft, the former deputy director of the Forest Service's fire and aviation program. "Politically, we have to fight fires. People see the smoke and demand to see helicopters and hotshots. And the Forest Service wants to send them in. This is an agency run by people who were raised fighting fires. They like it. It pays more than office work. It's more fun, and there's a clear case of good versus evil. They get to play hero."

Croft tells a story about a wildfire burning in wet leaves on Long Island one fall in the late nineties.

"It was out. There was no threat at all, but the public wanted to see somebody fight it. I got a call from the governor himself. He told me, 'I want to see some *fucking* air tankers.' So I put in a call to our base in South Carolina, they loaded up one tanker with a bellyful of slurry, and he flew up, dropped it on what was at that point a pile of smoldering leaves, the crowds cheered, and the fire was just as done as it had been before the air tanker was sent."

The tanker flight cost around $4,000.

"Wildfires are the only natural disaster we think we can control," said Chuck Womack, who runs the dispatch center at the NIFC in Boise. It's not a rational notion. "If the federal government gave us all the money in the world, we'd still fail to control all the blazes."

Federal agencies don't have unlimited budgets to fight fires, but they do spend a lot of money keeping flames away from houses. As recently as 1991, stamping out backcountry fires took up just 13 percent of the Forest Service's annual budget. Today the agency, the biggest in the wildland fire business, has assumed the role of preventing wildfires that start on publicly held lands from crossing boundaries into municipalities or privately owned property. Fighting fires now consumes nearly half of the Forest Service's annual budget, which most years approaches or exceeds $5 billion. Every year since 1999, the agency has overspent its suppression appropriations and has had to borrow millions from its other programs—timber, recreation, fisheries—to meet the need. Though Congress has reimbursed the Forest Service for up to 80 percent of the fire program's overspending, calling it disaster relief, the cannibalism of other program budgets has become so bad that it amounts to an identity crisis for the agency: Is the Forest Service's primary purpose fire suppression or managing the lands it oversees? The way things stand, there doesn't seem to be enough money for it to do both.

With this question unresolved, fire seasons are expected to grow 50 to 142 percent larger by 2050, and the population expansion is likely to keep pace. The Forest Service predicts that by 2030, 40 percent more homes will be in the path of wildfires. Right now, with federal, state, and local government spending included, one study puts the total annual cost of the grand experiment to control the flames at $4.7 billion, and there's little reason to believe that that figure will do anything but rise.

Yet there's no evidence that the increased spending is doing much to make towns that abut the forest safer. The flames are natural; the homes aren't. And with denser forest, drier climates, and more people living in the wildlands, wildfires are burning houses with a frequency never seen before. In the 1960s, just a hundred homes went up in smoke every fire season; today the number is close to three thousand.

Of course, these costs don't compare to those from a massive hurricane or tornado—Katrina cost $125 billion. The difference is that the threat wildfire poses to houses and towns can be mitigated—through forest thinning, prescribed burns, and defensible-space work. Yet western towns remain inexplicably ill-prepared for lurking catastrophe. In 2013, fewer than 2 percent of America's communities had done any defensible-space work at all. One retired incident commander says it's a small miracle when a year passes without an entire town or city burning. The miracle seasons are becoming rare, though. In back-to-back years, a pair of wildfires burned more than three hundred homes in the city of Colorado Springs.

Willis and PAWUIC's critical contribution to Prescott was that they recognized impending disaster and took action to mitigate it. At first, the community wasn't motivated to allocate sufficient funds. If there wasn't smoke, there wasn't fire to worry about.

With little local support, PAWUIC did what it could, like buying chippers for disposing of the trees and excess brush on the edge of town and spreading the motto "Living on the Edge" to convey the importance of proactively protecting homes from wildfires. As soon as PAWUIC drummed up enough federal and state grant money, it hired Crew 7 and two full-time employees to identify the most threatened parts of the city. Still, the town hall meetings PAWUIC held once a year to encourage homeowners to preemptively help fireproof their own homes went almost unattended.

The 2002 fire changed everything. Some eight hundred people showed up for the meeting after the blaze. By the time the second Doce Fire struck, more than two hundred thousand people lived in Yavapai County, and Willis's program was recognized as the country's "gold standard" by the National Fire Protection Association. The city was the first in the state to adopt a wildland-urban interface code: No new houses could be built without meeting code. Since Marsh and Crew 7 started their defensible-space work, they'd successfully protected eighteen thousand homes and more than $3.1 billion in property. At least on paper, there was no American city better prepared for wildfires than Prescott.

CHAPTER 15

HOMETOWN PRIDE

Donut picked a new theme song for Alpha. Chris MacKenzie was back as Alpha's lead firefighter, and Donut, once again in the back of the buggy, reclaimed his role as deejay. The tune was Rammstein's "Du Hast," a thrashing metal song that riffs on traditional German wedding vows and repeats "until death separates" ad nauseam. Nobody in the buggies, of course, knew what the foreign lyrics meant. What they liked was how the pulsing bass line psyched them up for the task at hand. Not long after noon, the hotshots arrived on scene. The wind blew the smoke north across Iron Springs Road, and the column was lying almost parallel to the ground. As they drove through the plume, the hotshots on the left side of the trucks could see a line of flames ten feet tall, maybe higher, churning through the brush and coming straight at the highway.

Donut turned around to the other hotshots in Alpha. "Get your fucking shit together," he said. "This is what we signed up for."

Steed rearranged Alpha and Bravo after Bunch's and Renan's departures, and Granite Mountain was a slightly different crew than they had been on Thompson Ridge. Marsh, who was getting his truck fixed and not with the hotshots when the Doce broke, was back in his superintendent's position on Granite Mountain. Steed stepped back to captain, and Tom Cooley returned to structural firefighting.

Clayton Whitted had moved Sean Misner, a twenty-six-year-old rookie from California with a new baby on the way, to Alpha to be Grant's new "battle buddy." Scott Norris and Joe Thurston shifted to Bravo to replace Bunch. John Percin, who had hurt his knee during the first week of the season, wasn't currently in Alpha but was expected to be back soon. Since his injury, Percin had been a "chip bitch," doing defensible-space work with the three other guys Marsh and Steed hired each year to keep the projects going while the crew was on the line. That morning, when Steed asked Percin how he was doing, Percin said, "Better, but I'm not a hundred percent yet." After two weeks on Thompson Ridge, few of the guys felt 100 percent, and with Renan and Bunch gone, the crew was now shorthanded and needed Percin back as soon as possible.

The buggies took a left down a dirt road toward the fire's starting point, where a spark from a bullet had ignited the chaparral beside a rock fence set up to keep ATVs out of the desert. The blaze burned so hot it looked as if the chaparral's stems had been blackened and then shorn with a razor, but the flames were already long gone. There was no chance the crew could cut direct line around the fire: It had aligned with the wind, and with the gentle rise in the terrain, the fire's head had already grown to a few hundred yards across. The firefighters would have to link together a series of roads to contain the Doce.

Granite Mountain backtracked on the dirt road, which was now stacked with arriving fire engines. Every available firefighter in the area had been pressed into duty. The plume was throwing embers across the road, and engine crews hurriedly hosed down the flames before the fire could get established on the wrong side of the impromptu containment line. Overhead, a single-engine air tanker strafed the unburned side of the road with retardant.

The roads between the flames and town were the firefighters' last hope for keeping the Doce from knocking on the doorsteps of the town's sixty thousand homes. If it crossed either road, flames would be running wild through ranchland with nothing between them and

Williamson—and, by extension, Prescott—but a few hundred heads of longhorn cattle and a five-mile strip of junipers and tangled brush that hadn't burned since the last Doce, twenty-three years earlier. To widen the line, Steed quickly planned to burn out a patch of chaparral that grew in the corner where the dirt road intersected the paved Iron Springs Road. With the flames spreading so quickly, it was a risky assignment. Steed needed experienced hands to pull it off.

"Chris, Donut—I want you to get into the interior and dot-light every twenty-five feet," Steed said. "This stuff is flashy. We just need to blacken the corner before the flames hit."

Chris and Donut would enter the chaparral alone and, parallel to Iron Springs Road, spark a small fire every twenty-five feet. The rest of the hotshots would spread out along the road to watch for spots.

The two men grabbed the torches from Steed's truck and waded into head-high thickets of brush. They could see the flames chugging toward the highway from the south. The boughs raked across their helmets, and they held their arms across their faces to block the whiplash of branches. Chris, the more experienced of the two, went deeper into the thickets than Donut. Forty feet off the highway, he came to a barbed-wire fence that ran parallel to the road. He pushed it down, stepped over it, and kept going, while Donut stopped at the fence, to stagger their burning positions.

"You ready?" Chris yelled to Donut.

"Affirm!"

They lit their torches in brush so thick it blocked their view of the road. Pretty quickly, Donut recognized the impossibility of the task. As he plowed through the brush, the burning end of his torch kept accidentally touching the chaparral. He was igniting the brush not every twenty-five feet, but with every other step. The tiny fires climbed down the branches, spreading from limb to limb until, within a matter of moments, the entire oak shrub was engulfed in flames and spreading in the adjacent brush.

"This is nuts!" Donut yelled. "These spots are right on my ass!"

"We gotta get the fuck out of here!" Chris yelled back.

They used their gloved hands to snuff out their torches and, just minutes after wading into the brush field, started back out toward the highway. But in the short time Chris and Donut had been burning, the Doce had accelerated, and the wall of flames now pushed even harder toward the road. Smoke was steaming through the chaparral. As Donut crashed through the brush, he could hear the popping of sparks, the roar of fast-approaching flames, and the superheated water in the plants whistling like a teakettle as it vented from the shrubs' woody stems.

Chris caught up to Donut and slammed into his back, and they both tumbled to the ground. On their hands and knees, they scurried over shotgun shells and empty Coors bottles and through the openings near the foot of the brush. Still, they weren't closing the distance to the highway fast enough. Here in the dense brush, their shelters would do nothing. If the fire caught them, they'd be severely burned or, more likely, worse.

"Throw the torches!" Chris yelled, hoping to lighten their loads. They heaved the gas-filled canisters up and over the brush and kept going.

By now they were nearing the highway, and the hotshots spread out along the road watching for spots turned to see Chris and Donut stumble from the brush and onto the safety of the blacktop. Behind them, the fire overtook the drip torches and the canisters exploded. The main blaze, meanwhile, was bucking into Iron Springs. The flames curled over the road, and the fire kept marching directly at town.

By evening, smoke filled the Prescott suburb of Williamson. Flames threatened more than two hundred homes on the outskirts of town. A Type 1 incident management team was brought in immediately, and within just a few hours of its start, the Doce became the highest-priority fire in the country. Tony Sciacca, a former hotshot superintendent who lived in Prescott, was the incident commander. By 5 P.M., Sciacca had assumed control of the Doce and was scouting the fire from the passenger seat of a small helicopter. He could scarcely believe what he saw.

The Doce now spread across more than five thousand acres, an area that, for reference, took the Thompson Ridge Fire four days to burn. Spot fires, thrown more than a half-mile ahead of the main blaze, had spread flames nearly three miles beyond the ignition point. Fire ran all the way over the top of 7,600-foot Granite Mountain and was pushing into the subdivision of Sundown Acres. Families, convinced their houses were going to burn, started stashing valuables in boxes and loading them into cars, while air tankers dropped thousands of gallons of red slurry on their backyards and local engines raced into the threatened subdivision against the flow of fleeing evacuees. Protecting Sundown Acres was triage.

An Incident Command Post was set up at Prescott High School, and the SWCC in Albuquerque ordered personnel and resources from all around the country, including the Forest Service's only two DC-10 supertankers—converted commercial aircraft capable of dropping an eleven-thousand-gallon payload that covers a three-hundred-foot-wide, quarter-mile-long strip. In one shift on the Doce, specially trained firefighters working out of the Phoenix-Mesa Gateway airport reloaded the DC-10s with retardant, and the planes dropped a million dollars' worth of slurry on the fire—so much that the red lines were visible from space with the naked eye.

This was exactly the kind of "adjustment in normal tactics and strategies" that Chuck Maxwell had said would be needed to contain fires in such fierce drought conditions. So far, it seemed to be working. By morning, just a few fingers of black reached between the houses in Sundown Acres; no homes had burned.

"We aren't out of the woods yet," Sciacca warned during a press conference. He knew that, at best, the slurry would check the Doce's spread, but if the winds returned, the retardant wouldn't hold back the blaze. Sciacca didn't feel that his management team would have a solid handle on the Doce until hotshot crews and bulldozers lined the entire fire perimeter. Accomplishing that task required building fifteen miles of line; Sciacca expected it to take seven to ten days.

Fourteen hotshot crews were ordered to build the line. With the possible exception of the Forest Service's Prescott Hotshots, none knew the area better than Granite Mountain. It was for this reason that the men were given a special assignment on the fire's far western edge. Out there, in the bottom of a dry and still-unburned canyon, there stood a two-thousand-year-old alligator juniper—among the world's oldest and largest specimens.

"That's the kind of loss you can't quantify," a Forest Service biologist had argued to Sciacca's management team. Granite Mountain drew the assignment of protecting the tree.

The tree was less than twenty miles as the crow flies from their station, but for more than an hour the buggies bounced down dirt roads that passed beneath Santa Fe Railroad trestles built in the 1880s, herds of longhorn cattle grazing among cholla and teddy bear cactus, and the massive white skeletons of drought-killed cottonwoods. Officially, this was called Division B.

Back in high school, the hotshots who grew up in Prescott had come out here to party. Since then, Donut had visited the spot periodically to shoot guns in the pullouts and makeshift ranges that spring up beside road signs, but he'd never seen the juniper.

The buggies followed a series of increasingly smaller dirt roads until the drivers parked where the terrain was too hummocky for any vehicle but a dirt bike. For thirty minutes, the hotshots hiked down a gently sloping arroyo. The forest air felt so dry, it seemed to crackle as if clicking with countless cicadas. Near the intersection of the wash and a rolling canyon, the hotshots saw the giant's canopy rising a few dozen feet above the other trees in the forest. But what set the tree apart more than its height was its sweeping, anvil-shaped top. Its trunk was fourteen feet in diameter. Donut thought it looked like something out of *The Lion King*.

An alligator juniper's thick and scaled bark looks like the skin of the reptile it's named after. Over time, the giant tree had become frayed from weathering, and great strips of dried bark hung from the branches. Many of its enormous limbs were long dead, a defensive mechanism that allowed the tree to save water by allocating it only to the most productive branches. Now the system that had kept the tree alive for millennia threatened to kill it. The old juniper was a great woodpile, its green canopy stacked atop dead branches that were effectively cured lumber.

Granite Mountain didn't have much time to save the tree. Behind it, a wall of knee-high flames was backing down the canyon's gentle flanks. One squad carved a circle out of the brush surrounding the ancient juniper while the other put in a quick piece of line just wide enough to burn off.

Steed put Bob in charge of burning out the brush around the tree and assigned two hotshots to help him, then led the rest of the men back to the buggies at a fast clip. Bob gave Steed and the others a few minutes to get out ahead, and in those few moments, Bob and the others climbed into the tree's limbs and snapped photos to document what they knew could be the last time anybody saw the ancient juniper standing. Then they set fire to the slopes and hiked out the arroyo as if racing a lit fuse.

A few days after their hurried burnout, Steed and the crew anxiously hiked back into the canyon to check on the tree. The slopes behind the juniper burned hot. When the brush, now gnarled and black, went up in flames, it had almost certainly sent a flurry of embers drifting upward. How the wind happened to be swirling in the canyon when Bob lit the backfire had determined whether the embers bent back into the juniper or, as the hotshots hoped, were swept northward in the general direction of the prevailing wind and the main fire. As it happened, the winds probably did both.

"Got a duffer up here!" Steed said, referring to a barely smoldering ember that sat in the teacup-size hole it had burned into the wood. Steed was straddling a branch eight feet up in the tree, and he extinguished the last remaining hot spot on the juniper. The tree still stood, but the heat had shriveled and curled back nearly a third of the green branches nearest the burnout.

Nobody could know whether the ancient juniper had the strength to recover from the stress of the wildfire. Some, like Donut, thought it would surely die. How any being, especially one so old, could survive such a severe burn was beyond him. Other hotshots weren't so convinced, but collectively, their mood was ebullient. If nothing else, they'd given the juniper a fighting chance.

For a short while, the men spread out around the tree and mopped up any remaining spot fires in the vicinity, but few remained. The chaparral had little wood left to harbor a spark. After, the men sprawled out in the shade of the juniper's branches and, while eating lunch, made deliberately crass jokes about visiting the tree during firewood-collection season.

Steed called for a crew photo after the men ate. Traditionally, Granite Mountain's crew shots were taken during their most exotic tours or after their hardest shifts. One was on a California beach not long after receiving their hotshot status; another was during the fire burning mysteriously in the rains of Minnesota's Boundary Waters; and a third after wrestling into submission an unexpectedly dangerous blaze on Arizona's Mogollon Rim. Clayton framed the photos with wood scraps and salvaged materials, and each picture hung proudly on the station walls. Over the course of a season, most hotshots spent hours lingering before the photos, studying the faces of men who came before them. Each image hinted at a crew's particular legacy. There, before the wall, they heard stories from veterans about a certain sawyer, or the fastest swamper the crew had ever seen, or some crazy one-year wonder who took a swing at a bouncer. Perusing the pictures of past hotshots made each man wonder what would be said of him when he left.

The hotshots fanned out into the tree. Wade and Clayton scooted out on some of the taller limbs and hung from them, grinning and swinging like orangutans, while the others leaned against a branch, as thick as most trees, that jutted out parallel to the ground. Donut, in a pair of aviator sunglasses, crossed his arms and adopted a look and posture that said hardened hotshot. Scott held his hands in front of him—quiet and respectful—and Grant squared up to the camera and gave a smug grin.

After the photo was shot, Steed yelled "Pyramid!"—one of the impromptu team-building exercises, like the theater moment on Thompson Ridge, that he sometimes liked to spring on the crew. Along with four of the bigger hotshots, Steed got down on all fours and, in decreasing order of size, the men climbed atop one another. The structure collapsed a few times, the men laughing in a jumble, before Grant was able to climb up the other men's backs to the second-to-last of the five tiers. Beside him was William Warneke, a twenty-five-year-old ex-Marine from California who had served a tour in Iraq. Billy, a rookie on Granite Mountain, had a wife and a baby on the way. With Billy and Grant in place, Woyjeck scampered up the juniper and lowered himself onto their backs—the top of the pyramid. Chris snapped a quick photo. In it, nobody is posturing, but everyone is smiling.

Granite Mountain's saving of the juniper became the biggest story to come out of the Doce. To the few locals who had hiked into the desert to see the enormous tree, the crew's act felt personally important. But for most people, something else was at work. Many townspeople knew of the Forest Service's Prescott Hotshots, across town, but didn't know that the city employed its own hotshots until the Doce—until Granite Mountain had protected an ancient tree on the flanks of its namesake peak. It was a great story. The Prescott *Daily Courier* ran an article about Granite Mountain's efforts and talked to Marsh about what it was like to have saved the iconic tree.

"It feels right, you know?" Marsh told the reporter. He hadn't been there when the crew saved the juniper. During the first stages of the initial attack, the incident commander had tapped Marsh to oversee a division of the fire, and once again he hadn't worked with Granite Mountain at all on the Doce. But he understood that the newspaper story wasn't just about the tree.

The hotshots' work often goes unappreciated, since it's usually done so far away from towns, Marsh noted. After so many years of working to protect Prescott, the recognition was well earned.

Granite Mountain was *news*. The men joked nonstop about being anointed heroes. In a show of thanks, locals brought to the station sleeping bags, doughnuts, muffins, and Gatorade. How any of Prescott's residents ascertained that the unmarked compound on an industrial street corner was Granite Mountain's station, the guys couldn't figure out. None of the public had ever stopped by before.

Another perk of fighting fire in their hometown was that the hotshots, unlike the out-of-town crews, could skip the chaos and discomfort of a crowded fire camp. After the juniper, the hotshots spent five more shifts improving indirect dozer line miles away from the main fire. Running chainsaw in the heat made for long, hot, and physically taxing days. After their days on the line, the hotshots slept at the station. Steed and Marsh wouldn't let the hotshots leave—they were technically on assignment—but they let the men's families visit every evening.

The station had the atmosphere of a carnival. Leah came the first night.

"What do you want me to bring?!" Leah texted Grant.

"Beer!"

But Grant knew Leah couldn't bring beer, and not just because she wasn't twenty-one. Decades earlier, hotshots could drink on fire assignments, but that ended in the 1960s when a hotshot crew got drunk on a fire in Oregon and rioted, throwing chairs and tables out the windows of a scenic train they'd taken on the way home from an assignment.

Marsh and Steed wouldn't let Granite Mountain drink at the station. So Leah brought Grant his backpack with speakers in it and his preferred vice: candy. When she got there, Grant proudly introduced her to his crewmates. She met his new battle buddy, Sean Misner, and his wife, Amanda, and Clayton and his wife, Kristi—even Anthony Rose, who had been the hardest on Grant at the beginning of the year. Whatever tension existed between the hotshots at the start of the season was gone. After the flurry of introductions, Grant and Leah sat alone together in the parking lot.

He winced in pain as he took off his boots. His feet were literally rotting. An angry shade of red spread across the tops and bottoms of his feet. Grant had been changing his socks regularly and had even tried some of those weird toe gloves, in which each individual toe sits in its own cloth sleeve, but still, the long hours in his boots was taking a toll. Leah was washing Grant's feet with warm, soapy water when Chris MacKenzie came up behind.

"You're not really helping him with that?" he asked.

"I'm spraying him down," said Leah, her hands working between his toes.

"You're absolutely doing that," Chris said, and walked away gagging.

Grant smiled at Leah. He no longer felt like an outsider. He kept comparing fighting the Doce to winning the home game in high school football. For Grant, the newspaper coverage made real the often intangible fact that, occasionally, their normally remote job had real results for people in towns. He was warming to the idea of being a hotshot.

Grant missed Renan, but in his absence, he also felt like he was making more friends on the crew. During a break in one of the long shifts building indirect line, Grant told Scott Norris about Leah's grandmother. Her ninetieth birthday was coming up. She'd compiled a life list, and every birthday she made it her goal to check off a bold objective that year. She'd gone skydiving at eighty-six and asked to smoke marijuana at eighty-eight. Grant had helped with the latter—in front of all the guests at her birthday party, he presented her with a rolled joint. Now Grant wanted to top that for her ninetieth. He wanted to get her arrested.

"Heather's a cop!" Scott said. "She'd do it for sure." And so they schemed up a way to pretend to arrest Leah's grandmother. Leah loved the idea.

Oftentimes, the men parked their trucks in a row beside the station, and in the evenings, some of the men chose to sleep in the backs of their pickups instead of on the ground. One night Clayton and Zup had happened to park next to each other. Clayton and Kristi were quietly talking in his truck when Zup stuck his head through her window and said, "Let me know if you guys want any privacy, if you know what I mean."

Kristi rolled her eyes. Zup let out a laugh and then jumped into the back of his pickup, just three feet away. For Kristi there was something priceless about being with Clayton at the station. She saw him operating in the world through which he defined himself. Watching Clayton and the easy way he had as a leader satisfied a curiosity in her. It gave her pride. For many of the couples, the evenings at the station felt stolen.

A few nights later, as the crew returned from another long shift, Grant texted Leah and asked her to bring his ukulele to the station. She thought it an odd request. Grant was tone-deaf and couldn't play a chord. But Leah brought it anyway, and Zup grabbed it from her when she arrived. He tuned the little instrument and sat on the concrete and began slowly picking his way toward a rhythm.

After some time, a song emerged, and Zup ad-libbed nonsensical lyrics about the Doce and the hotshots. The men, their wives or girlfriends, and a half-dozen kids gathered in a loose circle around Zup. Soon the other men started contributing made-up lyrics, and everybody laughed because it was contagious and fun. Clayton grabbed Kristi's hands.

"*Shhh,*" Clay said. "Papa Bear"—his nickname for Marsh—"is sleeping. We don't want to wake him up."

They stepped into the center of the group and started dancing. Grant followed immediately after. He pulled Leah into the circle, and there, in the station's parking lot, with their friends and families surrounding them, the couples danced beneath the off-colored glow of the sunset filtered through drift smoke.

After the other hotshots' girlfriends and families left the station, the men laid their sleeping bags down around the compound. Scott put his by the buggy barn, near the gate, and dozed until sometime around midnight. Then his phone vibrated and he woke quietly, took his beanie and placed it on his pillow, and filled out his sleeping bag with extra clothes to make it look like he was still in there. Then he crept out of the station gate, the way lit dimly by the streetlights.

Heather Kennedy never made it to the station during those evenings after the Doce. She couldn't; she worked the late shift. But during any free moments she could find, she and Scott texted with a frequency that bordered on neurosis.

Heather, still wearing her uniform, had parked in a dark section of an alley behind the station. Scott slid into the passenger seat beside her, and they kissed for a long time.

After Thompson Ridge, whatever fears Scott had about losing Heather had been put to rest. They headed back to their apartment in Prescott Valley, twenty minutes to the north.

In rapid-fire stories, Scott told Heather everything about the Doce, even things she probably didn't need to know, like the time Donut, who was afraid of snakes and refused to head alone into the bush, "took a shit in front of the whole crew, just right there beside the line." Scott told her about the plan for Leah's grandmother, and they talked about the articles in the paper. He didn't understand why everyone was making such a big deal about Granite Mountain's actions.

"Do you think of yourself as a hero?" Scott asked Heather. They'd had this conversation before. "I don't think of myself as a hero at all. We're glorified landscapers."

At home, while Heather got changed, Scott showered. They met in the kitchen, where he fell quiet and thoughtful. He'd been doing that around Heather a lot lately, pausing in the middle of conversations for what seemed like minutes. A few nights earlier, he'd stammered out, "I . . . I . . . I've never been so happy in my life. I'm so in love with you." They'd been talking about buying a house together, and the word "marriage" found its way into their conversations with increasing frequency. But this time he had something far less tender on his mind. He looked afraid.

Scott told her the story of Donut and Chris crawling out of the thicket a few days earlier, the twenty-foot flames whistling and crackling behind them. He wasn't sure they knew just how close they'd come. "It was awful," he said, and leaned on the kitchen island by the bookshelf. "In seven years of firefighting, I've never seen anybody so close to being burned over."

Heather and Scott sometimes talked about the dangers of their respective jobs. They'd even discussed what to do if the unthinkable happened and one of them was killed at work. Heather promised him that if Scott died, she'd break the news to his sister and parents. If Heather died, he'd do the same. But the chance that somebody would pull a gun on her during a routine traffic stop seemed more likely than Scott—or anybody on his crew—getting caught by a fire.

"It was close, but the last time hotshots—guys who really knew what they were doing—got burned was in 1994," he told her. "Storm King." She knew the story. He'd told it to her before at an IHOP restaurant, drawing on the back of a napkin to illustrate what happened. He'd penciled in the fire line through thickets of oak scrub, the winds of an approaching cold front, the spot fire that had started below the firefighters, the ridgetop where one smoke jumper outran the blaze, and the places where fourteen firefighters who couldn't had died.

For those chilling moments on the Doce, Scott had seen the same fate befalling Donut and Chris. Heather smoothed her hands down his forearm; he'd run nine tanks of gas through his chainsaw that day, and his muscles were still as taut as a steel cable.

"Storm King was nearly twenty years ago," Scott said. "It happens, but it's really rare."

Then they went to bed, and he and Heather lay wrapped in each other's arms until his alarm went off at 4:30 A.M. On the way out to the car, he dropped a wet washcloth in the dust by the front door, then wiped it across his face to cover the evidence of his illicit cleanliness.

After seven shifts on the Doce, the incident management team released Granite Mountain on June 25. The Doce wouldn't be declared 100 percent contained until early July, but it never again ran like it had in that first terrifying shift. Sciacca's well-coordinated initial attack had helped end the Doce quickly. So had the flashy way chaparral burns: Unlike the timber on Thompson Ridge, which could smolder for weeks, chaparral burns as if soaked in white gas, but only when conditions align. When they don't, brush fires tend to be relatively tame. On the Doce, the winds never returned.

The hotshots spent the next two shifts at the station. The workday returned to normal nine-to-fives, which in one very important way felt like days off: After work, all the hotshots could go home and do as they pleased.

Kevin Woyjeck moved in with Grant and Leah. The last week of June was the first opportunity since Thompson Ridge that Kevin had time to actually move his stuff. A night later, Grant, Leah, Kevin, and their fourth housemate celebrated his arrival by heading to a steak house for dinner and margaritas. After a couple of strong drinks, Kevin bounced around in the backseat on the drive home, singing the chorus to Daft Punk's "Get Lucky"—"She's up all night to the sun, I'm up all night to get some, she's up all night for good fun, I'm up all night to get lucky."

He called it his song of the summer, an ode to his inability to get laid. Grant, also buzzed, howled with laughter in the front seat while Leah drove, chuckling. Living together, though they'd done it for only two days, had drawn Kevin and Grant closer. It helped that Kevin was a good housemate. He was tidy, motivated, and funny. Leah kept pointing out that Grant, whose hyper-cleanliness alienated some of his previous roommates, had finally found somebody other than her that he could live with.

The crew spent the couple of station days they had after the Doce taking care of minutiae overlooked in the hustle between fires. Scott and the sawyers cleaned their chainsaws, replaced fuel filters, and put an edge on their backlog of dulled chains. Meanwhile, Grant, Woyjeck, and the other rookies removed trash bags of empty water and Gatorade bottles from buggies and sharpened their hand tools. New boxes of MREs were loaded into the buggies, the drip torch fuel, bar oil, and gas were all replenished, and the hotshots were ready for their next assignment.

While the rest of the crew used the string of down days to rest up, Scott Norris remained exhausted. He'd been staying up all night to see Heather. One day, he fell asleep on the bench press in the weight room and Donut, seizing the opportunity, tied him to the bench with flagging while the other hotshots snickered in the background.

Marsh returned full-time to the superintendent position that week, too. One of the first things he did was recap for the guys what had happened on the Dude Fire. June 26 marked that fatal blaze's twenty-third anniversary, and many of the environmental factors that had led to that catastrophe were once again aligning over the Southwest.

Chuck Maxwell, over in Albuquerque, and other fuels and weather experts had issued a litany of dire warnings over the past ten days. First came the Fuels and Fire Behavior Advisory that warned of fire growth and behavior more extreme than anything seen in the past fifty years. The Doce had confirmed Maxwell's predictions, and since then the weather had only gotten hotter and drier.

The National Weather Service issued an Excessive Heat Watch for Yavapai County for the last week of June. Near-record high temperatures of 105 to 117 degrees were predicted through the weekend of June 28–30, with an increasing chance of thunderstorms. Monsoonal moisture was sweeping into Arizona. As the leading edge of the atmospheric tide crept closer to Prescott, the humidity ticked up. Rain was coming. Arizona's fire season was nearing its end, but lightning would precede the moisture. So would high winds associated with the thunderstorms.

Marsh's message to the crew was to not get lulled into a false sense of security. He'd heard from Steed that the hotshots had proven themselves on the Doce Fire and on Thompson Ridge, but it was not by chance that the Dude Fire fatalities had coincided with the onset of the monsoon. Historically, the Southwest fire season's last gasp was incredibly dangerous. Dry fuels and hot days, plus thunderstorms and strong winds, produced a volatile mix of conditions—particularly in the chaparral.

So the hotshots kept up their station chores—Woyjeck cleaning the bathroom, Grant sweeping the floors—waiting for a fire to break and soaking in the slow shifts in the meantime. On one such relaxed evening, Renan Packer stopped by the station. Since getting back from Thompson Ridge, he'd been lying low. Per doctor's orders, he'd stayed in bed for much of the past two weeks. Before heading over, he'd bought chewing tobacco for the men and candy for Grant, but somehow he managed to forget the gifts at home. No bother, Renan thought. He'd drop them by later.

At the station, after catching up with a few of the guys, Renan met with the overhead. He looked frail, but he was walking. During the ride home from New Mexico, Marsh had offered Renan the chance to come back to the fuels crew. Renan had said he'd think on it, but he simply couldn't stomach the idea of working for Granite Mountain without actually fighting fire. Chipping brush was not why he'd signed up to be a hotshot. He didn't share these thoughts with Marsh and Steed, though. Instead he told them that he needed more time to recover.

As they sat in Marsh's office, Steed, Clayton, Bob, and Travis Carter talked for a long time about his intention of returning to the hotshots in the future. Renan maintained that the muscle cramps were an unfortunate fluke. The doctors said it was a back spasm unrelated to his health issues in high school. The condition, spurred by dehydration and exhaustion, could befall any hotshot. Renan wanted back on the fire line. He asked for a recommendation for the city firefighting jobs he planned to apply for in the meantime. Clayton readily agreed.

Eventually, Renan stood and shook hands with Marsh and the squad bosses and followed Steed into the saw shop, where the men had already gathered in anticipation of being sent home for the night. The hotshots gathered around Renan, slapping his back and wishing him well. Steed gave the floor to Renan.

"I'm not going to be coming back, and I'm not going to be working for the chip crew, either," Renan said. "I need the time to regain my strength. I want to thank you guys"—his voice cracked and his eyes welled. "Working here was the most fun I've ever had."

At which point Steed stepped in and stopped him from choking up too much. "Tunnel!" he yelled, calling for another one of his rituals. The men instinctually paired up and made a long triangle with their bodies. As Renan ducked and ran under their arms, the hotshots smacked his backside. It was childish, a relic from high school football teams, but for Renan it was poignant.

The guys broke apart and headed to their cars. Some grabbed Renan's shoulder for a moment to say goodbye, and others pounded his back hard—*We'll see you around, dude. Take care of yourself—no more spasms, you hear?*

Renan walked with Grant to his Dodge Neon. Renan updated his battle buddy on his condition, and Grant told Renan stories about the Doce. They laughed as Grant told him about the theater skits Renan had missed on Thompson Ridge.

"All right, brother," Renan said. "I gotta get out of here. Love you, man."

CHAPTER 16

FIRE ON THE MOUNTAIN

ois Ferrell, a seventy-four-year-old retiree who lived in Glen Ilah, a subdivision in the 650-person town of Yarnell, Arizona, saw the lightning strike. It was Friday, June 28. She was gardening by the koi pond in the backyard, watching the thunderstorm flicker over the six-thousand-foot peaks of the Weaver range, to the north.

The sky flashed and rumbled every few minutes. Some fifteen thousand feet above Glen Ilah, the flat base of the cumulonimbus clouds seemed to be boiling. Ice crystals and dust caught in the clouds' chaos rubbed against one another, creating an electrical field. Not unlike the glow of static electricity seen when freshly washed laundry is pulled apart, the flashes twitched and jumped between the clouds.

All at once, the negatively charged particles in the clouds drew a filament of positively charged particles upward from a brush-covered granite outcrop on the long ridge north and west of Lois's house. In a hundredth of a second, a stream of negative particles raced down from the clouds toward their opposite. Where the opposing charges met, the circuit closed. The electricity flashed earthward and the hair-thin bridge between sky and ground pulsed as temperatures surrounding the lightning bolt reached fifty thousand degrees—five times hotter than the surface of the sun. Just as fast as it came, the bridge broke and the lightning snapped back into the clouds. The supersonic boom followed, rattling the glass panes of the windows behind Lois.

A short while later, a wisp of blue smoke twisted up from the ridgeline. When Lois saw it, she called to her seventy-three-year-old husband, Truman, who was inside watching TV. A retired Air Force mechanic with a full head of gray hair, Truman had been the town's interim volunteer fire chief until two years earlier. The smoke caught his attention, but he figured professional firefighters would take care of it before too long.

By evening, Russ Shumate, an incident commander for the Yavapai County Fire Department, had taken up a scouting position across the street from Glen Ilah at the Ranch House Restaurant, a small-town diner with a sweeping view of the surrounding mountains. He scanned the hills for smoke.

After sixteen years battling fires in Yavapai County, he knew that dry lightning in late June usually sparked not one fire but dozens. Already there were nine burning around Prescott. Shumate expected more to appear in the drier peaks of the Weavers.

He called the Prescott dispatch and requested that an Air Attack plane fly over Yarnell. These small prop planes orbit fires and carry one-to-three-person teams of highly trained firefighters who, in addition to scouting the fire's spread and helping incident managers plan the operations, also coordinate the aerial attack. Air Attacks serve as lead planes that guide in large air tankers, help helicopters pinpoint their bucket drops to the places they're needed most, and orchestrate the movement of all of these resources to avoid a midair collision. Along with a strong tolerance for airsickness (turbulence beside smoke columns can be violent), the job requires responding to and keeping track of radio traffic that pours in from as many as nine different radio frequencies.

It was late evening when Shumate heard an Air Attack plane whining over the Yarnell area, and by dark, the aerial firefighter and Shumate had located four new fires in the hills outside town. Truman was right about the one Lois had spotted on the ridge to the west. It was half an acre max, burning in a boulder field, with one active corner, low spread potential, and no structures or people at risk. Still, Shumate wanted it out. Late June was the wrong time of year to let brush-fires burn. But it was so small that Shumate determined that a hotshot crew wasn't needed to handle the fire. Instead he put in an order for an engine, two inmate crews, and a light helicopter to be on scene early next morning. And he gave the fire a name: Yarnell Hill.

A day after the fire started, on the evening of the 29th, Marsh was having dinner with his wife, Amanda, at the Prescott Brewing Company when the local dispatch called. The crew needed to be at the station first thing in the morning to pick up any more fires that had sprung up in the area. With all the new starts, it wasn't yet clear which fire was going to be the most pressing, but Yarnell Hill had become a problem.

That afternoon, the winds kicked up and the fire began to threaten Yarnell and Peeples Valley, a smaller ranching community to the north. Shumate upped his resource order considerably from the day before. Now there were fourteen engines, two structure-protection specialists, air tankers, three hotshot crews, two bulldozers, and a Type 2 incident management team assigned to handle the increasingly complex fire. Marsh didn't know any of this yet.

"Just another lightning bust" is what the dispatcher had told Marsh, and what Marsh told Amanda. Always the southern gentleman, Marsh excused himself after getting the call and notified Steed, who was at home drinking a Coors Light on his porch and watching Caden and Cambria swing in the backyard. Steed called Clayton and Bob, who called the men of their squads while Marsh, a painstakingly slow eater, settled back down and worked his way through his meal.

Prescott felt sultry that Saturday night. The pulse of moisture that had brought the dry lightning storms to Yarnell had pushed the humidity into the twenties for the past few days. After dinner, Marsh and Amanda stepped outside to enjoy the arrival of the monsoon and the onset of the Southwest's most pleasant weather. They walked hand in hand under the elms surrounding the courthouse, past children racing around a Rough Rider statue and young couples flirting on the park benches. Whiskey Row ran along one side of the square. Inside Moctezuma's, one of six bars lining the street, Brandon Bunch's dad poured drinks for Donut, Chris, and Zup. They'd stopped in for a beer and claimed seats in the back of the bar by the pool table and TVs replaying Wimbledon and baseball highlights.

"Why the fuck would you wear that?" Zup said to Donut. He was talking about his friend's pink tank top, which looked straight out of Southern California. "Respect yourself."

"What are you doing out of your cage?" Donut came back. It was good to be back out with the guys. "Shouldn't you be home listening to vinyl records with your girlfriend?"

Donut had come down with a nasty cold the morning before and hadn't worked the past two days. The father of a close friend of his had passed away, and Donut had spent that morning at the funeral before hanging out in front of the TV with Chris's poorly behaved dog, Abbey, a cattle dog with a big personality that Chris refused to discipline. The apartment felt empty without Chris; they'd been inseparable since he got back from California, and without his friend around, Donut felt restless and lonely. TV and video games could do only so much to fill the void. Heading out with the guys felt good.

"So what's the plan for tomorrow?" Donut asked. "Anybody know?"

"Another lightning fire, probably," Chris said.

The hotshots had spent all of Friday evening cutting line around a twenty-two-acre blaze near a mountaintop northwest of Prescott, one of the other small blazes that had broken out after the storm. They slept in a campground that night and woke up on the 29th with another small fire to deal with in the chaparral outside Skull Valley, about halfway between Yarnell and Prescott. That blaze was awful: no shade, little action, and nothing but sweat, smoke, and a growing irritability among the guys from lack of sleep. During that shift, Scott backed into a cactus and spent the afternoon pulling spines out of his ass. It was both annoying and a little humiliating, but it wasn't as bad as the time a skunk had sprayed him. Really, Donut hadn't missed much over the past few days. The biggest news was that after a couple of months chipping brush, John Percin had finally returned to the crew from his knee injury.

Percin had come to Prescott from a Portland, Oregon, suburb, joining Granite Mountain after a stint in rehab and a job waiting tables at a Mexican restaurant. While the rest of Granite Mountain fought New Mexico's Thompson Ridge Fire, Percin had stayed home doing inglorious fuels work in Prescott. Then he missed the Doce and saving the juniper. When he returned to the hotshots, he'd promptly gone down with heat exhaustion on the Skull Valley fire.

But other than Percin's unfortunate introduction to crew life, Donut hadn't missed more than a few hundred bucks of overtime. Zup and Chris didn't know where the hotshots were headed in the morning, nor were they very concerned about it. Whatever assignment they drew meant, most important, more overtime. Granite Mountain had taken only two days off since the Hart Fire, and June was shaping up to be a profitable month for the crew. They'd already racked up 261 hours of overtime that year: more than $3,000 for each seasonal hotshot.

At Heather's place outside of town, Scott Norris couldn't sleep. Heather was working graveyard. To pass the time alone, Scott played with their puppy, watched TV, showered, and did the dishes and the laundry. He paced and resigned himself to the fact that he wouldn't see her. He wasn't pleased.

"Fuck! Oh well, maybe I'll see you tomorrow night. Hope for no more fires," he'd texted her earlier that day, after the small Skull Valley fire and before he got the call from Steed. Sunday should have been Granite Mountain's day off.

This was how it had gone all week. One night, Heather had decided to bail on a mountain of paperwork to see her boyfriend when she had the chance. It was early morning by the time she made it home. Another night she snuck out and they took a shower together. He'd pulled out the cinnamon gum he was chewing and placed it atop the stall. Heather gave him a look.

"I'll take care of it later," he said. "I promise."

Marsh had decided to sleep at the station. It was more convenient than making the hour-long round-trip drive home to the ranch. When the men started trickling in at around 5 A.M., he was up already, drinking his fancy coffee. He'd brewed a cup from the Jetboil he carried—along with the poet Gretel Ehrlich's book *In the Empire of Ice*—in his line-gear pack.

Since Marsh had now reclaimed the superintendent position, June 30 marked the first time since before Thompson Ridge that Granite Mountain had a full roster of twenty hotshots. They pulled up the folding chairs in the ready room. They kicked their freshly oiled boots on the tables and filled the time with talk of the night before—food, beer, sex, sleep—while waiting for the others to arrive and Marsh to begin the morning briefing. On one of the whiteboards behind him were the daily factoids—"Cows that listen to music make more milk"—and on the other were their "Daily Physical Percentages." Some of the guys had tried to estimate what they had in the tank—Chris MacKenzie was at 77 percent, Scott was 74, Marsh was 68—while others turned it into a joke: Ashcraft was "'Stache-less" and Grant was "10 squared," while Donut was simply "Hell Ya."

Marsh was no doubt pleased to be back with the crew. After so many years on the line, the varied and compelling rhythm of firefighting was easy to miss, but his time away had been good for his career. While Steed and the guys were out chasing fires in New Mexico, Marsh had been dealing with the city's defensible-space planning and making future project work for Granite Mountain when the summer fires slowed.

Amanda never believed her husband could handle the time away from the line. Hotshotting, she thought, was too much a part of Eric's personality. But maybe the bigger question was whether he could handle giving up the crew he'd built. Relinquishing control of Granite Mountain was far more personal than simply finding a suitable replacement, which Steed had proven he was.

Transferring power to Steed seemed more promising than it had with Aaron Lawson, the Forest Service captain Marsh never got along with. Steed agreed with the Prescott Way, understood Marsh, and was ex-military. The Marines ingrained in their soldiers the importance of following orders. And over the 2012 fire season, Steed and Marsh had made a good team, fighting fires together for more than 120 days.

But in the summer of 2013, Marsh and Steed had hardly worked together at all, and in many ways Steed had already taken over the crew. Marsh must have seen how the men coalesced around his younger colleague. Wade Parker called Steed his "hero." Donut and all the other hotshots agreed that he was a "Greek god." The veterans found Steed more easygoing and fun than Marsh, and he was the only superintendent that the rookies and new guys like Scott knew. To them, Marsh remained an enigma.

Even so, if Steed was in any way reluctant or disappointed to hand the crew back to Marsh, he didn't show it. By the time all of the hotshots arrived at the station, Marsh knew Granite Mountain was heading to Yarnell. He took his place at the head of the ready room. The young men quieted down immediately.

"We're going to a lightning fire. It's at four hundred acres and threatening some houses in Peeples Valley," Marsh said, as terse as ever. "Load up."

CHAPTER 17

INITIAL ATTACK

The forty-six-mile section of Highway 89 between Prescott and Yarnell was built in the 1920s, and it's still rough enough to make even the road-worn carsick. It took more than an hour for Granite Mountain to travel south, up and over the Bradshaws, through the ponderosas outside Prescott, and into the heavy chaparral brush and beavertail cactus that covers the Weavers around Yarnell. Bob texted his wife, Claire, "So much for days off. Heading to a 500-acre fire in Yarnell. Love you."

Scott texted his mom. He'd broken dinner plans with his parents the night before for another chance to see Heather. "Sorry I didn't make it last night. I guess I was tired and fell asleep early. We've been reassigned to Yarnell Hill ☹. Not very excited about it."

Then he texted Heather: "I love you baby. Have a good weekend. I'll try and keep you informed if I have service."

Sometime around 6:30 A.M., the hotshots pulled into Yarnell. Built in 1888 during a brief gold-mining boom in the Weavers, the town has forever since tried to reinvent itself as a tourist town. Its slogan: "Where the Desert Breeze Meets the Mountain Air." But tourism never got the economy going. If Yarnell, at forty-eight hundred feet, is cooler than Phoenix, it's also warmer than Prescott. The town's greatest draw is a toss-up between the Shrine of St. Joseph, a Jesuit collection of marble Jesus statues, and the biscuits and gravy at the Ranch House Restaurant. There's no supermarket, no gas station, and one bar that holds odd and inconsistent hours. Most days of the week, the Ranch House and two side-by-side antiques stores, one of which proudly displays the town's logo, a rising buzzard, are the only places open for business. Their customers are most often aging Harley riders or weathered locals in jeans and cowboy hats. Still, the 650 retirees, artists, and eccentrics who call themselves Yarnellers love their home. The people here look after one another.

Geographically, the area around Yarnell is as complex as it is striking. Granite peaks rise to the east and west of town, and scattered everywhere are enormous boulders eroded into the twisted shapes of a Dalí painting. The town is perched on the very southern edge of the Colorado Plateau. Just a few hundred yards beyond the Ranch House, Highway 89 drops eighteen hundred feet over nine miles down to the Sonoran Desert, among the hottest and driest places in North America. At that precipice, one can look out over two hundred miles of desert wilderness. Watching the sunrise from the top of Yarnell Hill is often spectacular, especially during mornings when monsoonal moisture and a little smoke refract the light.

The 30th was such a morning, but Granite Mountain didn't catch the sunrise from the Ranch House. They stopped a mile north at the volunteer fire station where Shumate, the incident commander, had been running the firefight from an informal Incident Command Post. Except for a few 4×4 trucks from the state and county that were backed into parking spots, the fire station was almost entirely empty. Granite Mountain was the first crew to arrive that morning. The field generals Shumate had called up from around the state were trickling into the station. Those who had arrived already were clutching cups of coffee while organizing the day's firefight. To a man, they were gray-haired and had spent their careers on dangerous fires.

There was Byron Kimball, the fire-behavior analyst who studied the fuels and weather to predict how the fire would spread. Paul Musser, the former superintendent of the Flagstaff Hotshots, would serve as one of the two operations chiefs, along with Todd Abel, a firefighter from Yavapai County who had fought fires for eighteen years near his Prescott home. While Musser commandeered additional resources, Abel would direct the on-the-ground tactics of the firefight. Gary Cordes, a mustached Yavapai County fire chief, was the structure-protection specialist for the town of Yarnell.

The fire was still effectively an initial attack. On June 30, Shumate was no closer to catching Yarnell Hill than he had been when he arrived in town on the 28th. But Shumate now looked worn thin from exhaustion. He hadn't slept in thirty-plus hours, and he wouldn't be able to until his replacement, a baldheaded and red-faced man named Roy Hall, arrived on scene at around 10 A.M. to take over as incident commander. There was a lot of work to do before then.

The men gathered around a table, and Shumate spread out a spiral-bound map picked up at the local minimart—perplexingly, the best the growing incident management team had—and shared what he knew about the fire. Thus far, its storyline was short, but its potential was explosive. The fire was burning on top of a ridgeline three miles west of town. Shumate had initially deemed it too far to hike to, and the day before he had ordered a BLM helicopter to fly six firefighters from an Arizona Department of Corrections crew to the ridge. For most of the 29th, the fire crept through the understory of brush thickets, but then the wind picked up late in the afternoon and the fire size jumped from a half-acre to dozens of acres as the flames started running through the chaparral.

Shumate ordered a heavy air tanker and helicopter to keep the blaze in check, but both pilots declined the mission because strong thunderstorms had developed over the runway. So Shumate stuck with the single-engine air tankers—SEATs—available at two nearby municipal airports. By the night of the 29th, pilots had dropped more than seven thousand gallons of retardant along the fire's flanks. This did little, and the blaze only got more volatile, at one point burning over a cache of Gatorade and extra fuel flown in to resupply the inmate crews.

By 6 A.M. on the 30th, Yarnell Hill approached three hundred acres and burned in the shape of a cone. Its pointed end started on the ridgetop and spread half a mile to the north. One flank hung up on the rimrock along the long crescent-shaped ridge that ran between Yarnell and Peeples Valley, and the other had rolled down into the valley at its foot. The fire's head sat just a mile away from Double Bar A Ranch, the first seven of the forty-two homes in the Model Creek subdivision, which was soon to be impinged upon. The gathered overhead knew the fire could close the distance before six that night.

The firefighters agreed that Model Creek and Peeples Valley, a dispersed community of 428, four miles north of Yarnell, were the first priority. They'd dedicate most of the few dozen incoming resources to doing point protection—building a moat of defensible space around the ranch and letting the fire burn around it. A network of roads and isolated horse ranches in Peeples Valley gave them a decent shot at protecting the community. If the fire became increasingly unmanageable, they could use the roads for lines and intentionally burn off the lighter and flashy fuel in the pastures. Protecting Yarnell, though, was a much different situation.

"My assessment is if this fire hits Yarnell, we're going to lose the town," Gary Cordes, the structure-protection specialist, told the others.

He'd been scouting the area since the night before and was using an iPad and Google Earth to study the ground. The house-size rocks that added to Yarnell's charms were now problems. Many of the homes were tucked into the boulders, so their back doors effectively opened to stone walls. Thick brush surrounded the homes and ran nearly without break to the fire's edge. Well out of the jurisdiction of the Prescott Fire Department and the city's progressive wildland fire policies, the homes in Yarnell, like those in most towns in the West, had almost no defensible space.

"Yarnell's poorly positioned, and the vegetation is very decadent," said Cordes. The area hadn't burned in forty-five years, and beneath the chaparral were kindling piles of long-dead limbs and dried leaves.

Byron Kimball, the fire-behavior analyst, was worried, too. He later compared the brush to a kiln-dried two-by-four. Fuel moisture, or the percentage of water in flammable materials, is the metric used to determine how ready vegetation is to burn. The fuel-moisture content of a two-by-four bought new at Home Depot is usually 10 to 12 percent.

"That same two-by-four, had it been lying on the ground at Yarnell Hill, would probably be four percent fuel moisture," Kimball said. Fuel did not get drier than that. From very fine grass to the centuries-old junipers that grew thick near Peeples Valley, Kimball's message was that all that vegetation was ready to burn. The men discussed the best course of action for saving Yarnell, and they didn't like their options.

The one upside was that they had a bit of time. With the daily winds pushing the fire's head north toward Peeples Valley, Yarnell's five hundred homes were threatened primarily by the slowly burning flank. Barring a wind shift, Shumate estimated that the flames wouldn't get to Yarnell until the morning of July 1.

Cordes, the structure-protection specialist, suggested that the best option for protecting Yarnell lay in a two-mile-long overgrown road that connected the town to the ridge where the fire burned. If the fire continued north toward Peeples Valley, as Cordes expected it would, then pivoted with a wind shift and ran at the town, as everybody knew was possible, firefighters should have plenty of time to burn off the road. Everyone agreed to the plan. They ordered a bulldozer to widen the road into a reliable fuel break.

Then Cordes added something else. *If the fire does make a run at Yarnell,* he said, tapping the iPad, *go here—the Helms' place.* The twenty-to-thirty-acre Boulder Springs Ranch sat about a quarter-mile west of Glen Ilah, at the mouth of an east-facing canyon in the Weavers. The roof was metal, the siding was metal, and there were no bushes or trees flanking the barn or the main house. It was just about the only place in Yarnell with adequate defensible space.

"It's a good safety zone," Cordes said. "Absolutely bombproof."

Eventually the Management Team broke into smaller groups and Abel, the tactical operations chief, stepped outside with Marsh.

They'd worked fires together for the better part of a decade, and Abel had even written a positive review of Marsh's performance on a previous blaze. A big and gentle man who preferred bow-hunting elk and antelope to most things in life, Abel wore a cowboy hat and sported a handlebar mustache. Out front of the station, they could see the Weaver Mountains and a bubble of blue smoke puffing up from atop the ridgeline.

Marsh asked Abel if it was anchored. The line around the heel of the fire needed to be started from something nonflammable—the rimrock or the cold black—to keep the flames from outflanking firefighters.

"I don't think it is," Abel said.

Anchoring the fire seemed like a good place to start. Abel asked Marsh if he'd be Division Alpha and oversee the western side of the blaze and anchor the fire. Making Marsh Division Alpha compartmentalized responsibility into a chain of command. Rather than having to run the crew, Marsh would take a step back, focus on the best way to contain the western side of the fire, and help coordinate the greater effort. Though Marsh and Steed would be working more or less side by side, officially Steed would have command of the hotshots while Marsh would be Abel's point of contact and control all the resources in his division. At that point, that was only Granite Mountain.

It was a choice assignment for Marsh. He was tasked with executing Abel's plans, but he didn't have to answer to anybody. As division, he had the latitude to refuse assignments and choose whatever tactics best accomplished his goals on the piece of ground he oversaw. Essentially, Marsh was free to run Division Alpha as he saw fit.

"That ranch I was telling you about, the Helms' place, is over in that area," Cordes said, pointing off to his left. He stood with Marsh on the edge of Yarnell, where the houses transitioned to the brush. Neither of them could see the ranch through the boulders and a wall of chaparral nearly as tall as Marsh's truck, but he got the idea: There was an obvious safety zone beneath the long ridge of the Weavers. "And of course, you've always got the black," Cordes added.

He'd led Marsh and the hotshots through Glen Ilah, at the southern end of town. Mature oak and elm trees shaded the subdivision of a few hundred ranch-style homes, none of which looked defensible. Earlier that morning, Cordes had driven to the fire's edge on the convoluted network of jeep trails lacing the area—the only access. He explained that the two-track Marsh and Granite Mountain wanted led to a rusting and long-abandoned road grader at the foot of the mountains before turning left and climbing 850 feet up to the ridgetop. There, at the fire's heel, Marsh and Granite Mountain could start the task of anchoring the fire.

Granite Mountain drove until the brush on both sides of the trucks became thick enough to scratch the paint. The buggies parked in a small clearing in the chaparral, and Steed had the hotshots begin gearing up for an hour-long hike to the fire. Meanwhile, Marsh, with his smaller truck, went a half-mile closer and headed in on foot to flag a route for the men to follow. From a distance, the hotshots could see Marsh's red helmet bobbing as he climbed up toward the ridge.

At that point, the fire was still calm. There were a few fingers of blackened fuel left behind by embers that had rolled downhill and ignited the chaparral in a few subtle drainages and draws that flowed toward the valley below. Every so often, a bush along the fire's perimeter would flare up, but the blaze wasn't progressing much. Wisps of smoke hung in the air like a few hundred individual campfires lit on the hill.

Shortly after 8 A.M., Marsh reached a saddle on the ridge below the fire's edge, and encountered a man and a woman from the nearby town of Congress, who had hiked up to see the blaze.

"What are you doing with them pink ribbons?" the woman asked him. Marsh, who must have been surprised to see hikers so close to the fire, explained that the flagging was to mark a path in for the crew, and asked what the couple were doing up there.

They'd come up the mountain to see the fire and take pictures for a friend of theirs, a cashier at the local gas station, they said. It quickly became clear to Marsh that they'd spent a lot of time in the area. The couple told him that they'd started hiking at Glen Ilah, bushwhacking through brush so thick "a bear couldn't roll in it" to a thin road that ran parallel to the ridgetop. They told Marsh that if they had to do it again, they'd skip the bushwhacking and just walk in on the road that ran from the fire's edge, along the ridgeline, straight back to Glen Ilah, and almost directly to the Helms' place. The hikers left the mountain shortly after.

June 30 was the most chaotic day of the summer for Chuck Maxwell and the dispatchers at the SWCC in Albuquerque. The Southwest was ablaze. The same wave of thunderstorms that sparked Yarnell Hill had sent more than nine hundred other lightning strikes crashing into Arizona's and New Mexico's mountain ranges. Almost thirty fires were burning across the region, and Maxwell was predicting record high temperatures and more dry thunderstorms. Nearly every incident commander wanted additional resources, and most of those orders were funneled through Albuquerque's command center. Phones rang constantly as dispatchers tried to track down more equipment.

That morning at 8:30, the commanders on Yarnell had ordered a DC-10 and two more large air tankers. The nearest plane was in Albuquerque, but it hadn't made it as far as Yarnell. Another fire that had started the day before outside the twenty-eight-thousand-person town of Kingman, Arizona, was growing quickly. Evacuations would soon be in effect. The Kingman fire was deemed a priority, and the SWCC diverted the plane, as well as the second DC-10, to the blaze.

Maxwell, who constantly kept National Weather Service colleagues apprised about fires around the region, sent out an email update: "Some resource allocation chaos right now. There are more orders than there are resources. One of the non-environmental factors that supports large fire growth and activity!"

The environmental factors were far more troubling. A layer of moisture that originated in the Gulf of Mexico and the Gulf of California was tracking toward Prescott and Yarnell. In that high band of atmospheric moisture was enough water to support the growth of scattered thunderstorms. As the day grew warmer, the water would condense into small cotton-ball clouds that would eventually hang up on mountain ranges, where they would develop into thunderstorms before drifting south and southwest with the general atmospheric movement prevailing over the region. At each mountain range the storms passed over—the San Francisco Peaks, the Black Hills, the Bradshaws—they would increase in volatility. Maxwell had no way of knowing for certain where the thunderstorms would develop, or if they would even reach as far south as the Weaver Mountains and Yarnell Hill, but the potential existed, and if the storms did hit, they'd come during the heat of the day, when the fire would be burning at its hottest.

Between the dryness of the fuels and the presence of potentially strong thunderstorms, June 30 was turning into just the type of day he'd forecast for the Prescott area months earlier. All of the environmental factors were aligning. Maxwell broadcast his forecasts to the region's firefighters, along with a warning most had internalized since their first season on the line: *If you see thunderstorms, expect strong, erratic winds.*

Darrell Willis, the structure-protection supervisor of Peeples Valley and the Wildland Division chief for Prescott, had come to Yarnell after Shumate called him in the middle of the night. The incident commander desperately needed seasoned help. Willis, with his decades of experience, could see that homes were being threatened when he pulled into Peeples Valley at 3 A.M.

By 10 A.M. the Yarnell Hill Fire was becoming agitated. The fire now approached fifteen hundred acres. A flaming front nearly a mile and a half wide stretched across the valley and had taken the arched shape of a parachute catching the winds blowing off the desert floor. It was outrageously early in the morning for the blaze to be burning in chaparral as if it were the heat of day—something Willis had never seen in thirty-five years of fighting fire. Forty-to-fifty-foot flames churned through the brush, burning north at a quarter-mile an hour. From this point on, all the fire behavior could do was intensify.

The seven structures in Double Bar A Ranch were under the most immediate threat. Willis assigned a Department of Corrections hand-crew and a Forest Service engine to protect the ranch while two other engines started thinning the brush around the thirty-five other homes in the Model Creek subdivision. But even with the advantage of roads and pastureland to burn off, Peeples Valley was going to present some serious challenges to the firefighters.

Double Bar A had wood-shake roofs and zero defensible space. A caretaker gave Willis and the firefighters access to a ninety-thousand-gallon water tank, and they set up sprinkler systems to promptly spill most of it on and around the houses. Sawyers began clearing the junipers and chaparral on the side of the ranch facing the fire as fast as they could cut. But creating a solid border of defensible space around the ranch might take most of a day. They had only hours.

Everything looked different for Granite Mountain. Relative to the wall of flames roaring toward Willis and Peeples Valley, the fire's back edge was calm. Marsh gave out assignments, briefed the leaders on safety zones (the black, the ranch) and escape routes (the ridge road), and went ahead to scout the ridgeline to the north. Donut's assignment was to anchor the fire's cold northern edge. He led a saw team up to the ridgeline. Steed and Bravo squad started lining the fire's more active eastern flank, a few hundred yards below Donut.

Though creating an anchor point was a critical assignment, the job wasn't a particularly dangerous one, just uncomfortable. The low brush covering the ridgetops provided no shade, and at 10:30 the temperature was already past ninety. Before hiking, Steed had ordered the men not carrying saws to pack extra water because, at the fire's heel, there was no promise of a timely resupply. Like many of the guys, Donut had carried in thirteen quarts of water, adding twenty-six extra pounds to his pack. To compensate for the heat and the extra weight, Steed led the men at a slower pace than usual, even stopping for water breaks on the way in. Donut downed a quart and a half of water before reaching the ridgetop.

While the saw team cut a wide swath in the brush along the fire's edge, Donut hung back. The few remaining hot spots in the otherwise cold black lay near the stumps of the heaviest brush, and he used his Pulaski to splay open the little pockets of heat, exposing the coals to the breeze and letting the embers burn themselves out. The work was slow and tedious, but the stunning views made it easier. Little clouds cast shadows on the folds of brown mountain ranges and valleys that extended as far as Donut could see to the south, which was most of the way to Phoenix. In the valley where Yarnell sat, a pair of green buggies from the Forest Service's Blue Ridge Hotshots were parked by Granite Mountain's trucks. The crew, the second on scene, milled beside the buggies, looking neither pressed for time nor concerned about the fire, while the bulldozer cleared a safety zone around the vehicles. It was obvious to every firefighter on scene that the real action lay to the north, where the smoke column was building steadily.

Shortly before noon, the superintendent of the Blue Ridge Hotshots, Brian Frisby, and the crew's captain, Rogers "Trew" Brown, drove a Razor, a four-seat ATV, up to the ridge to meet face-to-face with Marsh and Steed. Blue Ridge's overhead wanted to ensure that everybody was on the same page. So far, it didn't seem like anybody was. Blue Ridge had arrived in Yarnell at 8 A.M. and still didn't have a clear idea whose division they were on or what the day's plan was. The firefight at Yarnell Hill still seemed disorganized.

The supes and captains shook hands on the saddle, leaning against their tools and looking over the fire in the valley while they compared notes and observations.

As he often did during slow moments on a fire, Marsh pulled out the fingernail clipper he carried in the front pocket of his yellow and started trimming his nails. He told Blue Ridge about the hikers and repeated the miscommunications he'd had with Air Attack around

9 A.M., shortly after first meeting with the hikers. When the crew first arrived at the black edge, Marsh had ordered Granite Mountain to burn off a corner of the two-track road they'd hiked in on, but the SEATs kept dropping retardant on his burnout. Air Attack had effectively disregarded Marsh's tactical choices. It wasn't a big deal—it meant that the hotshots had to cut direct line around the fire's heel—but Marsh was frustrated.

Another trouble was the radio frequency, a common issue on fires. Both crews were missing communiqués because of dead spots in radio coverage. To make contact with Abel, Musser, or Cordes, the captains and superintendents often had to switch between a line-of-sight tactical frequency and a command network that used a repeater set up in Yarnell to broadcast the radio signal over a wider area. Sometimes the repeater worked, sometimes it didn't. As a workaround, they'd started using firefighters in the valley as a human relay system.

Trew and Frisby noticed odd things, too. As Blue Ridge's overhead understood it, their assignment was to improve the dozer line that Cordes, Yarnell's structure-protection specialist, wanted built that morning. But the dozer operator now working in the valley wasn't fire-line qualified. Even his presence on the fire line was a breach of protocol. Trew, Blue Ridge's captain, gave the dozer operator an extra radio and assigned one of Blue Ridge's squad bosses the responsibility of keeping him safe. Then the dozer had tried to open up a safety zone around the old road grader that sat at the base of the ridge but stopped when he saw an alarming sign: DANGER! EXPLOSIVES. KEEP OUT! Maybe the explosives had been left there during an old gold-mining operation? Nobody knew, but the dozer operator, along with the firefighters he was working with, refused to use his tractor as a minesweeper.

"This is like the Swiss cheese effect," one of Blue Ridge's hotshots said to Trew.

"We'd need a piece of cheese for that," Trew replied. "This is like one big hole."

Then the chaotic scene got even more confusing. Marsh got a radio call from Division Zulu, somebody he hadn't heard of yet. Operations Chief Todd Abel had sent Rance Marquez, a BLM firefighter who had come up from Phoenix, to divide up control of the fire's western flank. Abel needed structure protection in Yarnell, and that demanded different tactics, and therefore a different division, from the pure wildland firefighting needed to control the fire's heel. But either Abel forgot to tell Marsh that Division Alpha was going to be split or Abel's message hadn't gotten through, because Marquez's radio transmission was the first Marsh had heard of a Division Zulu.

Marsh responded curtly, and several strained exchanges followed. The two knew each other from the Doce Fire, where Marquez had worked under Marsh's division. There, the men's relationship had been terse at best. To Cordes, who knew Marsh from years of working together in Yavapai County, the radio conversation at Yarnell sounded like a turf war that Marsh had no intention of losing.

Division Alpha's primary assignment—to contain the fire's heel— promised to be relatively uneventful. But if Marsh let Marquez divide Division Alpha into two pieces, he'd be left babysitting a dead piece of line on what was already proving to be one of the most dynamic fires of the 2013 season—if not his career. Maybe Marsh wanted to use his power as Division Alpha to get Granite Mountain a higher-action assignment, like burning out around Yarnell or cutting direct line somewhere near the head, or maybe he just didn't like the idea of being usurped by a firefighter who had been his subordinate only days earlier. Either way, Marsh fought to retain oversight of his portion of Yarnell.

"I'm not trying to take over your division," Marquez told Marsh multiple times over the radio. But, of course, he was doing exactly that.

Without settling anything with Marquez, Marsh called Air Attack to inform him of where he thought the divisions should split. The aerial commander sat in a high-powered prop plane, making wide circles above the fire.

"Go ahead, Division Alpha."

Compared with the chainsaws and the wind noise that muffled most ground resources' radio calls, Air Attack sounded as if he were transmitting from an NPR studio. Marsh explained to him that Division Alpha controlled Granite Mountain, Blue Ridge, and the dozer—all the resources on the Yarnell side of the fire except the engines working in town. Immediately after, Marquez got on the radio and told Air Attack the opposite: Zulu had Blue Ridge and the dozer.

Marsh snapped. "Listen. We need to decide something and go with it," he radioed out. It was typical of his transmissions: concise, curt, a little condescending. Zulu, clearly cowed, apologized over the radio, and the matter appeared to be settled.

Marsh called Todd Abel on the cell phone. "I wanted to call you and let you know that . . . well, I . . . I had to kind of get aggressive with this Division Zulu guy," he said. "I had to kind of be a little more assertive on the radio than I usually like."

"Well, did you sort it out?" Abel asked.

They had, kind of. Marsh retained control of Granite Mountain. Cordes controlled the dozer. Division Zulu wasn't heard from for the rest of the day. And Blue Ridge was still confused about who their division head was and what task they'd been assigned to do. The four groups of firefighters on the southern flank were in disarray.

CHAPTER 18

EYES IN THE GREEN

Donut finished connecting the line to the cold black and met up with Marsh and Steed for his subsequent orders. Granite Mountain's next step was to go direct by building line on the side of the blaze nearest Yarnell.

The flank sat about halfway between the ridgetop and the valley floor, and in such steep terrain, Granite Mountain would be taking a risk by cutting line across the middle of the slope. If a burning ember rolled downhill and ignited a spot fire below the hotshots, the flames could surprise the men by running uphill at them. That was one factor in the deaths of fourteen firefighters on 1994's Storm King Fire.

Steed and Marsh wanted a lookout to warn the crew about any spot fires so the men had time to step safely into the black. As superintendent, it was Steed's job to pick lookouts, but Marsh, likely unfamiliar with Steed running the crew, did it for him. Marsh picked Donut. After his sickness, a slow shift might help him recuperate. Steed didn't argue.

While the saw team tied back into the crew, Donut hiked the ridge, looking for a perch that allowed him to see past a pair of spur ridges that blocked his view of the slopes beneath the fire's flank. He couldn't find a spot he liked on the ridgeline. The place Donut liked was a half-mile from the saddle, on a knoll in the midst of the un-burned fuel in the valley. The knoll was just to the north of the old road grader with the EXPLOSIVES sign behind it. If it put him in a sea of green and unburned chaparral, it also offered an unbroken view of the slopes beneath the crew. As long as the winds remained favorable, Marsh agreed that it was a good place for his lookout.

"Watch out for that finger," Steed warned Donut before he left. "It's got potential."

As the fire's head ripped toward Double Bar A Ranch, a strap of flame had broken off and run up a ridgeline on the opposite side of the valley from Granite Mountain. That strap was now backing slowly south toward Yarnell. It wasn't much of a threat to the scrape and sawyers cutting line on the long ridgeline where the overhead now sat, but the finger could prove dangerous to Donut. If the winds did shift and the fire turned to the south, his lookout would sit right in the path of the flames. Donut, Marsh, and Steed took comfort from the small clearing that surrounded the old grader. From the ridge, it looked big enough to provide Donut with an ample safety zone.

Donut nodded his acknowledgment to Steed's warning, then pulled the extra water out of his pack and gave it to Steed. As a look-out, he wouldn't need it. He'd be monitoring the fire's progress, which meant sitting. Then Donut climbed into Blue Ridge's Razor with Trew and Frisby and rolled away from Granite Mountain.

"God, it's awful," Truman Ferrell said to his wife, Lois, the woman who had first spotted the fire two days earlier. Like many Yarnellers, the Ferrells had gone out to Highway 89 to join the groups of worried locals watching the blaze grow.

Truman, who had the fading tattoo of an anchor on his forearm, came to a vantage on the highway with his neighbors Dan Schroeder and Dorman Olson and their two Scottie dogs. They wanted to see if anybody's house was going to burn. Truman lit a Marlboro Light.

Truman had served in both Vietnam and Desert Storm. To him, Yarnell Hill looked like a war zone. Single-engine air tankers dive-bombed the fire. Helicopters dropped buckets of water on the flames. Local news crews were stacked up along the highway reporting on what looked to be an impending tragedy. At around 1 P.M., a reporter tweeted, "Yarnell Hill Fire forces evacuations of Model Creek subdiv. & Double A Bar Ranch area, shelter set up at Yavapai College in Prescott #fox10."

The whole town had smelled like woodsmoke for two days now, but Truman didn't become deeply concerned about the fire until Saturday afternoon. All night on the 29th, he could see the fire glowing on the ridge above town, and when he woke on the 30th, most of the ridge had turned black. Like many residents in Yarnell and Peeples Valley, Truman and Lois responded by packing up their lives' valuables in case of evacuation. Insurance documents, jewelry, the nine-millimeter pistol he'd gotten in Saudi Arabia during Operation Desert Storm—Truman and Lois stashed it all in the motor home and made plans for how they'd get out if it came to that. Truman would drive the Winnebago, towing their purple Honda CR-V behind, and Lois would take his pickup. Fortunately for the Ferrells, it didn't look as though evacuating Yarnell would be necessary.

"I think we're about to watch Peeples Valley burn," Truman said.

Not long after noon, few things could have sounded sweeter to Darrell Willis, who was still working at Double Bar A Ranch, than the low roar of the DC-10's jet engines coming down from the north. Albuquerque's SWCC had forced the fire outside of Kingman to release the aircraft. Yarnell Hill, having grown to more than fifteen hundred acres, now claimed the unfortunate title of the country's highest-priority blaze.

The fire leaped in bursts between the junipers and scrub oaks a half-mile from the homes in Peeples Valley. Small cyclones laced in ribbons of flame formed at the interface between the calm air and the upwafting smoke on the fire's edge. Like dust devils, they twisted into the unburned brush, igniting shrubs as they turned, dissipated, and formed again. The fire had already reached the first homes in Peeples Valley.

The DC-10 lumbered over a ridgeline off to Willis's right and, flying bizarrely slowly for such a massive aircraft, dropped to only a few dozen feet above the tops of the squat juniper trees. It laid a mile-long strip of slurry between Willis and the unburned chaparral. Shortly after, a second heavy air tanker reinforced the retardant line along a similar path. For the first time that day, Willis had a line between himself and the flames. But whatever relief that small safety net provided was short-lived.

Willis watched the fire collide with the first retardant line. For a long moment, the flames sat down, and it looked as though the slurry had broken the fire's advance. But the gusts of wind that came over the next few minutes breathed life back into the embers and, like a stumbling but still-dangerous boxer, the flames rose again and worked through the brush covered in red slurry. Once on the far side of the retardant line, the fire's intensity resumed as if it had encountered no retardant at all. Willis had run out of ways to stop the fire. Flames would be at Double Bar A within the hour and at the doorsteps of fifty or so other homes in Peeples Valley within two.

"Get him out of here," Willis said to the enginemen, referring to the caretaker at Double Bar A. The ranch hand, like many stubborn homeowners in Peeples Valley, had willfully ignored the evacuation orders already in effect, preferring instead to stay and do what he could to help the firefighters save the ranch. Even after Willis pointed to a pair of side-by-side tennis courts and told him that if the fire burned into the ranch, that was their last resort, the caretaker decided to stay. But his fortitude cracked after he saw the fire's intensity up close. With no interest in finding out what it felt like to ride out a firestorm on a pair of tennis courts, the man fled.

Shortly after, so did Willis and the firefighters he commanded.

"Everybody out! Just everybody get out of here," Willis screamed.

Scott texted Heather. "So this is how my morning's going. Structures threatened in Peeples Valley!"

She was at the vet, getting Riggs his rabies shot, when she got his text. The photo showed a ripping fire in the distance, but she couldn't see the homes through the smoke. Either way, Scott's assignment didn't look good. It looked hot. Scott hated hot. He wouldn't be happy.

Granite Mountain took lunch at a group of large boulders on the slope where they'd been cutting line. The fire still chugged along steadily to the north, and in the distance the men could see cauliflower clouds hung on the peaks of the Bradshaw Mountains outside Prescott. The clouds cast dark shadows over the distant hills. Drenched in sweat from running saw all morning, Scott cut open the rubbery packaging of his MRE lunch with his pocket knife and leaned back into his pack as the men around him traded for preferred food. *Pound Cake for Patriotic Cookies, anybody? Who doesn't want their Tabasco?*

Scott's phone vibrated. "I had a weird dream that I proposed to Scott." Then a few seconds later: "Oh hi. That was meant for Sarah. Lol."

Delighted, Scott thumbed out a reply to Heather.

"Well, that's better than the last one"—the last time, she'd dreamed that Scott was shoplifting. "I'm a little old fashioned. I think I'd like to be the one to propose. ☺"

"LOL. Okay. ;)" Heather answered.

Minutes later, the screaming of the DC-10's jet engines broke up Scott and Heather's digital moment. Scott looked up to see that the plane was on what looked like a collision course with a Type 1 helicopter, a Skycrane outfitted to carry water. The DC-10 crested the ridge, and the pilot must have seen the helicopter flying in from the northwest, but the jet couldn't change course. It was too big and already flying too slow to maneuver rapidly. The helicopter's pilot had to move, but its best option seemed to be an impossibly narrow draw that drained into the valley Donut sat in.

Donut, who'd arrived at his lookout nearly an hour earlier, was certain he was about to witness a midair collision. He had the wherewithal to pull his phone from his pocket and shoot video. Just a few hundred yards away, the helicopter pilot banked hard, then harder still. The Skycrane dived deep into the V-shaped valley, its nose still pointed directly toward the ground. Sixty, fifty, forty feet from the rocks.

I'm about to pick up a body, Donut thought. The helicopter's blades *thwapped* the air as the pilot pulled out of the maneuver and the DC-10 blasted overhead. Rotor wash kicked up plumes of dust as the helicopter pilot pulled the ship's belly parallel to the slope and re-gained control near the bottom of the gulch.

The DC-10 never altered its course. A moment after passing above the Skycrane, the pilot opened the bay doors and dropped an 11,119-gallon strip of retardant across the valley floor between Donut and the strap of flames that Steed had warned him about earlier that day. The safety buffer was comforting, but Donut was still shaken up.

On the ridge, Scott thumbed out another text to Heather.

"We just watched a DC-10 slurry bomber almost collide midair with a Sikorsky helicopter!"

"Holy hell! That certainly would have made the news," Heather replied.

Watching two multi-million-dollar aircraft nearly smack into each other stood out as ten seconds of terror in Donut's otherwise slow afternoon. He'd picked a trigger point—if the fire hit a drainage about a quarter-mile away, Donut would flee—and listened to Marsh and Steed chatter about the aircraft's near disaster and the chaos unfold-ing at the fire's head.

Donut ate his Beef Stew MRE for lunch, not bothering to use the chemical heater to warm the meal in the day's excessive heat. He sat just below the top of the blanched knoll. The winds blew at his back, where the long valley funneled the breeze rising off the desert floor to the north. But outside the valley's microcosm, the terrain shaped the winds differently. At around 1 P.M., the dominant breeze over the fire began shifting to the east, and after burning four of the seven build-ings at Double Bar A Ranch, the flames took a hard right. The fire's head now spread directly toward Highway 89 and the houses in Pee-ples Valley. Roy Hall's management team was already talking about shutting down the highway and using it as a line to burn off. None of the smaller residential roads seemed to be holding.

From the ridgetop, Marsh could see the fire's development, but in the valley, Donut couldn't. Donut's job was simple: In the event that a spot fire appeared beneath Granite Mountain's midslope line, he was to warn Steed and the crew. No spots had. As the crew ate their lunches on the ridge that formed the valley's northwestern boundary, Donut watched the strap of fire now backing toward Yarnell on the opposite ridge. The flank hadn't advanced more than fifty feet in the hour since Donut had been at his lookout, but the twitchy way the flames moved was fascinating. Rather than a steady creep, they throbbed in temperamental pulses.

One moment, the black edge was calm—almost no smoke at all. The next, the embers glowing at the trunk of a scrub oak warmed the lower branches, and in a flash of flame the whole bush combusted as if torched by a lighter. Then, just as quickly as the little conflagration had come, the flash smoldered. The effect was biblical.

As part of Donut's job to keep an eye on the fire, he "slung weather" at the top of every hour with a paperback-size kit every lookout carries in his or her line gear. At 1:50, he pulled out a thermometer on a small chain. Over the thermometer's bulb was a cloth wick. Donut dipped the cloth into his water bottle. Wetting the cloth provided both temperature and humidity readings. Then he held the end of the chain at arm's length and swung the thermometer in a circle for a timed minute.

Donut scratched the readings into a log used to keep track of hourly weather changes. "Cloud build up SW," he wrote. Out over the desert, Donut could see small, white puffs of clouds starting to pull together into thunderstorms. Then he finished the entry: "104 degrees, 10 percent humidity, 5–10 m.p.h. with gusts of 15 from the south."

CHAPTER 19

WIND SHIFT

Studying his radar in Albuquerque, Chuck Maxwell saw the wind shift coming long before anybody else did. A hundred-mile line of scattered thunderstorms had hung up on the Mogollon Rim and the Bradshaws. At shortly before 3 P.M., Maxwell watched a few of the clouds puff up like popcorn above a sea of white, a sign that the storms were reaching maturity.

"See that?" Maxwell asked a former hotshot who was now working as a dispatcher at the SWCC. The weatherman pointed to an eyebrow of shifting pixels on the southwest side of the thunderstorms looming over Prescott. "It's an outflow boundary. Exactly the stuff we teach in weather classes."

Rain now fell over the Bradshaws. Even if most of the raindrops evaporated in the fifteen thousand feet of hot and dry air that sat between the cloud's base and the ground—a phenomenon called virga—the moisture cooled the surrounding air and caused it to sink, pushing out a wave of winds. Meteorologists refer to this as an outflow boundary.

Maxwell was witnessing not the collapse of a single thunderstorm, but the collapse of many. What had been a line of mushrooming cumulus clouds, their tops reaching nearly thirty thousand feet, was now generating thirty-five- to forty-five-mile-an-hour winds. The clouds were pancaking out into a thin layer of overcast spreading to the south. Like tributaries strengthening a river, drainages and mountain arroyos funneled the winds created by the dying storms, and they eventually became a single gust front, now moving straight at Yarnell.

"This is going to get interesting quick," Maxwell said. "When it hits the fire, it'll turn back on itself and blow up. On the ground, there will be fire behavior that's about as extreme as it gets."

Maxwell would have been more worried had he not been in contact with Hall's management team all day, telling them directly that an outflow wind reversal was coming. His job wasn't to advise on tactics, but Maxwell's advice was implicit in the way he worded his warnings. When the winds hit, the firefighters and civilians needed to be well out of harm's way.

The dispatchers behind Maxwell fielded calls at an increasingly frantic pace. Incident commanders throughout the region knew of the coming storms, and to help mitigate the impending surge in fire behavior, they wanted more engines, hotshot crews, and, most pressingly, aircraft. In the past hour, the Albuquerque dispatchers had been inundated with requests: Yarnell Hill alone wanted six more large air tankers. The SWCC, already out of aircraft in the region, relayed the incident commander's request to Boise.

By that point in the afternoon, the State of Arizona had upgraded Yarnell Hill to a Type 1 incident, and new commanders were en route to relieve Hall and his team on July 1. In just a couple of days, the fire had gone from an errant lightning strike to the nation's highest-priority blaze. Still, when NIFC got Hall's management team's requests for more aircraft, they were promptly denied.

"Very limited availability of air tankers with increasing activity in the western states," came the response. "Unable to fill at this time."

Nine aircraft were already committed to Yarnell, and to provide Hall's team with any more air resources would mean risking disaster by pulling tankers and helicopters from the dozens of other quickly evolving fires around the West. NIFC couldn't do it. The planes couldn't be spared.

———

Around 3:30 P.M., **Donut** was slinging weather in preparation for his hourly report to Granite Mountain when another weather update came over the command frequency. The National Weather Service had announced the imminent wind shift. Storms building over Prescott had reached maturity.

Abel, the operations chief, radioed each unit under his command to confirm that they'd received the update. Division Alpha replied that it had.

"Have you got eyes on both of the cells?" Abel asked.

"Affirmative," Marsh said. Except for Air Attack, Marsh had the best view of the fire. Abel intended to use his vantage as an asset.

"Okay, watch that one to the north. It's making me nervous. It's collapsing and building," Abel said. "I watched it do it two or three times."

Granite Mountain was eating at its lunch spot when the weather update came through. Scott immediately texted Heather. "This fire is going to shit burning all over and expected +40 hr wind gusts from a t-storm outflow. Possibly going to burn some ranches and house."

When Heather got the text, she was at home refilling the dogs' water bowl. It was her day off, but she was catching up on the paperwork she'd abandoned the night before to spend time with Scott. The storms were already hitting Prescott. Outside, the winds blew so hard that she worried for a moment that the quaking windows might crack. They seemed to bend inward as the gusts buffeted the house. She grabbed her phone off the kitchen counter and texted Scott back, "I love you baby. Talk to you later!"

At 3:30, the flames pulled harder than ever northeast into Peeples Valley, and the brush was torching just a few hundred yards from the Incident Command Post, which had been moved to Model Creek Middle School, in Peeples Valley, to accommodate the incident's increasing size and complexity. The vehicles at the command post were pulled farther back into the safety zone, while the engine crews set fire to the last road between it and the flames.

The burnout wasn't an ideal plan. The backfire would almost certainly impinge on the already evacuated structures to the south, but the firefighters hoped the action would trade a few homes for many. Keeping the fire from jumping Model Creek Road would stop it from crossing into a subdivision of modular homes just past the safety zone at the middle school. The barter wasn't working. As soon as the torches were set to the ground, embers began sparking tiny fires on the north side of the line. Engine crews bailed into the brush and hosed down the spots, but there would soon be more flames than water.

The shift came in stages. The band of cooler, moist air—the outflow boundary—that had originated from the thunderstorms over Prescott moved south down the back side of the Bradshaws and across the arroyos and parched fields of Skull Valley. Like the small wake that breaks off a ship's bow, the front pushed forward the slowly moving air ahead of it. The weak waves sheared and angled as they hit pressure and temperature changes. When this confused front hit the bubble of hot air rising off the fire and the daily winds blowing off the desert, the air masses battled, creating an atmospheric instability that effectively canceled out the winds. A moment of calm fell over the area.

Rather than spreading aggressively forward, as the fire had been doing all day, the flames flickered and stood and worked their way through the already ignited brush. In the calm, the smoke column, which had been lying down to the northeast, was freed from the influence of the desert winds and billowed skyward. Firefighters stopped and looked at one another in disbelief. The calm was momentary and duplicitous, but it allowed the engine crews working on Model Creek Road to pick up spots threatening to overrun their line. The burnout, which had seemed destined to fail, turned and slowly blackened its way south, and the threat to the middle school and the Incident Command Post was nullified. Peeples Valley was spared.

But like a tide moving from north to south, the full force of the outflow boundary overpowered the desert winds. It first hit the flames on the blaze's north end, whipping the fire into a frenzy. Like a bloodhound picking up a new and stronger scent, the flames pivoted and ran hard to the south. Truman Ferrell was still on Highway 89 when he saw what had been the fire's smoldering southeast flank jump to life.

"Uh-oh," he said. "That don't look good. That's going right to Glen Ilah."

To keep himself entertained on the lonely knoll on the valley floor, Donut started spinning the thermometer on its chain fifteen minutes earlier than normal. That would give him time to check and double-check his readings. It was just before 3:30 when Steed called on the crew network.

"Donut, you up?"

"Go, Steed."

"There's a weather report coming in on the radio"—it was the second one in as many hours. "Spin your weather and listen to crew. I'll tell you what's going on."

Donut checked to make sure his radio was properly set up, then he went back to spinning the thermometer.

Marsh was among the first firefighters on Yarnell Hill to feel the shift. He radioed Todd Abel and told him that the "winds are getting squirrelly" on the ridgetop.

"Copy that. Are you in a good place?"

"Affirmative, in the black," Marsh said. "And trying to work my way off the top."

It wasn't clear if Marsh meant that he was trying to work his way off the summit of the Weavers, which was a half-mile or so to the north of Granite Mountain's position, or the ridge itself, which extended in a long arch back toward Yarnell and the Helms' place. For the moment, Marsh's position didn't concern Abel. He'd heard what he needed to: Granite Mountain was safely in the black.

"Okay, I copy, just keep me updated," Abel said to Marsh. "You guys, you know, hunker in and be safe, and we'll get some air support down there ASAP."

Meanwhile, Donut kept swinging the thermometer. He glanced back up at the ridge to see if the crew were still at their lunch spot, but they'd since left the fire's creeping edge and moved deeper into the black on the ridgetop. By the time Donut looked back to the fire, a ten-to-twelve-mile-per-hour wind was blowing into his face.

As Donut watched, the flames instantaneously jumped to life. What moments earlier had been a three-mile-long smoldering edge suddenly became a three-mile flaming front that was burning directly at Donut and the town of Yarnell. It was just what Steed had warned him about. Donut stopped spinning weather. Just seconds after the wind shifted, the fire started racing through the unburned chaparral on the valley floor. The flames jumped a drainage a quarter-mile to the north.

"Steed, Donut. It's hit my trigger point. I'm heading back to the safety zone."

As soon as Donut dropped off the back side of the knoll and started running toward the old road grader, he lost sight of the fire, but he knew it was taking off. The smoke leaned over Yarnell and billowed white and heavy off to his left. For the second time in as many weeks, Donut found himself in a footrace with flames.

The branches whipped his face, and Donut wove through the path choked by the fewest bushes. As he raced off the knoll, he considered his options. The thin road that led back up to Granite Mountain wasn't one. Fire burned fastest uphill, and with the crew being a half-mile away, the risk was too great to even consider. Same with the buggies parked in the flats a third of a mile from his lookout. On foot, there was no way he could beat the wind-driven flames to the buggies. That left two options: Call Blue Ridge's superintendent, Brian Frisby, and hope he was nearby with his ATV; or deploy his fire shelter. The thought of having to weather the storm in an aluminum tent produced a wave of nausea. A line of thirty-foot flames was sweeping toward him.

He broke out of the brush thicket into a world of disappointment. The clearing wasn't nearly as big as it looked from the ridge—it was a little larger than a tennis court. This wouldn't work as a safety zone; it may not even have been large enough for Donut to safely deploy his fire shelter, a reality that, regretfully, he had overlooked when he was dropped at the lookout hours earlier. The wind beat the thick walls of brush surrounding the grader.

"Donut, I can see you in the dozer push," Steed called. Even over the radio, Donut found it comforting to know that Steed and all the hotshots were watching him.

"I'm calling Blue Ridge," Donut said.

"Total nonstop chaos" is what Brian Frisby, the superintendent of the Blue Ridge Hotshots, called June 30. Highway 89 was packed with media and civilians. The radio traffic from the Peeples Valley end of the fire was a constant flurry of such-and-such compromised house or such-and-such compromised line, and the tactics seemed to be changing as quickly as the blaze itself. When the danger became too high, Frisby considered houses just another fuel type—not something worth risking lives to save—but he knew that the presence of homes sometimes prompted otherwise logical firefighters to make higher-risk decisions. Wildfires near towns were always the most complicated.

Blue Ridge was working on Yarnell's Shrine Road, near the marble statues of Jesus. The thin paved street ran perpendicular to Highway 89 and up Harper Canyon, a dry wash on the town's north end that was shaded by oak trees and packed with homes. The Shrine Road completed the contingency line the dozer had started building that morning. If the fire threatened Yarnell, Blue Ridge could keep the flames away from town by burning out, but the crew wasn't finished with the line and wouldn't be for some time.

Just before the weather update about the wind shift, Marsh had called Frisby requesting another face-to-face meeting. Things were changing quickly. He wanted to coordinate Blue Ridge's and Granite Mountain's efforts. Frisby agreed and took off in the Razor to meet with Marsh on the ridge for another talk. He never got there.

When he rounded a corner on the two-track, Frisby saw thirty-foot flames jumping from the fire's previously sleepy southern flank. *God dammit.* Granite Mountain's lookout was in serious trouble. Smoke cloaked the north side of the knob where he and Frisby had dropped Donut off earlier in the day. Frisby accelerated, bouncing the ATV over the rutted and dusty bulldozer tracks.

By the time he reached the little clearing, Donut was standing by the grader with his radio in his hand, looking stunned. He'd just keyed the mic to call Blue Ridge—the only way for Donut to be extracted from hell—when he saw that Frisby was already there. Frisby whipped the Razor into a U-turn. Donut handed him the radio and said, "Talk to Steed and Marsh." Donut wanted them to hear from Frisby that he was safe.

"I've got Donut and he's leaving his lookout," Frisby said. "The fire activity is picking up and we're moving our rigs. Do you want us to move yours?"

"Affirm," said Marsh. He watched as Frisby floored the ATV and they raced away ahead of the flames.

CHAPTER 20

EVACUATE IMMEDIATELY

All of Granite Mountain had heard the radio traffic about Donut getting blown off his lookout, and from the safety of the ridge they had seen the rescue. They could see Frisby and Donut in the Razor, a cone of dust billowing out behind as they outpaced the fire to the buggies. The wall of flames burning toward Yarnell now spread across the whole valley.

After receiving the weather update, Steed had pulled the hotshots off the direct piece of line they'd been cutting, and the men now sat close together on rock perches near the ridgetop. Ash drifted down on them. The mood was somber but not tense. Torched manzanita skeletons and the black and withered flesh of burned yuccas surrounded the hotshots for hundreds of yards in all directions. From where the men sat, the view was apocalyptic, but the black was cold and safe.

Corkscrews of dark smoke twisted off the leading edge of the fire and pulled into a column so thick it blotted out the sun. The smoke reached thirty thousand feet, as high as the cloud cover moving in from Prescott. Even the blue sky on the column's edges looked smudged, but the men kept their eyes glued to the valley-wide flaming front. It had very nearly reached the knob where Donut sat just moments before.

The hotshots kept their packs on, as if they were just pausing for a few moments before standing to move again. The four sawyers sat close together with their swampers nearby.

Dustin DeFord clipped his gloves to a pink carabiner on his shoulder strap and folded his chaps up to let out a bit of body heat. Zup did the same. At that point, every hotshot from Percin to Steed understood there was no longer reason to cut line. As Willis had witnessed that morning, the flames barely slowed at the retardant line the DC-10 had dropped: $10,000 worth of slurry rendered useless in moments. Another weather change was the only thing that could stop the Yarnell Hill Fire.

Grant sat below the sawyers. He leaned against a rock, rested his hands on the Kevlar chaps covering his thighs, and watched the flames race toward Yarnell. In a nod to his steady improvement throughout the season, Steed and Clayton had promoted Grant to swamper—an honor for any rookie, and recognition that he'd proven himself one of the fittest men on the crew. But that afternoon, Grant looked resigned. Over the past few hours, he'd seen aircraft nearly collide and Donut narrowly avoid a burnover. Now he was going to watch Yarnell burn. The day had started slow, but the afternoon was punctuated by brief adrenaline-filled moments. The experience was both exhausting and exciting. Yarnell Hill was news in the making, and Granite Mountain had front-row seats to a catastrophe. Grant, like the rest of the hotshots, couldn't look away from the fire.

Wade snapped a photo of Grant with his phone: the flames engulfing the valley on the left, Grant in the middle, and Yarnell on the right. Wade texted it to his mom, along with a note explaining that the town was soon to burn. The sawyer Andrew Ashcraft did the same, sending a photo to his mom and his wife.

Scott texted Heather. "Holy shit! This thing is running straight at Yarnell!" It was 3:52. Heather didn't respond.

Marsh wasn't with the crew while the hotshots rested. As was his style, he was likely scouting the long ridgeline closer to the Helms' place. The intensifying fire behavior might call for a change in tactics on Division Alpha, and Marsh always walked the ground he worked to familiarize himself with the terrain. As he hiked, he radioed Steed, who was resting on a flat rock with the hotshots. Bob, Turby, and Tony were sitting next to him.

"I was just saying—I was just saying I knew this was coming when I called you and asked what your comfort level was," Marsh told Steed. "I could just feel it, you know."

Marsh was stating the obvious—they'd received multiple weather updates forecasting the wind shift. That the fire would explode when the winds hit didn't require great foresight. Whether it was that or something else, a few veterans seized on Marsh's statement as an opportunity to poke fun at their boss.

"Yeah, well, we've been feeling it all day," Turby said, spitting out a long strand of tobacco. Tony and Chris chuckled. So did Steed. Bob, who seemed to be consumed with thought or worry, leaned on his knee and looked out over the fire.

"I'm just saying, you know. Too bad," Marsh continued.

Steed ignored Marsh's train of thought, and a long moment later he replied with an update on the location of the fire.

"It's almost made it to the single-track road that we walked in on," he said, referring to the road Donut and Frisby had used to flee minutes earlier.

As Donut and Frisby sped down the dusty track toward Granite Mountain's buggies, Frisby radioed Trew, his captain, and told him that they needed another driver to move all the vehicles. *I'll be there shortly to pick you up,* Frisby told Trew. The flat gray of overcast skies—an artificial sunset—supplanted the bright light of day as the smoke cloaked the sun. Flames would be at the buggies soon.

When they got to the trucks, Frisby told Donut to take off if it got too dangerous.

"But otherwise," Frisby said, "wait here until I get the other drivers."

He wanted to lead Donut out of the labyrinth of jeep trails that threaded through the valley. As Frisby sped off toward Blue Ridge, Donut, out of danger for the time being, threw his Pulaski and gear in the back of Marsh's superintendent truck. He climbed into the cab, turned on the A/C, and switched the radio to Granite Mountain's crew net.

On the radio, Donut could hear Marsh and Steed discussing whether the crew should stay in the black or come up with a plan to move. Marsh said he was scouting the escape route: the two-track that ran along the top of the ridge to Glen Ilah and the Helms' place. Staying put in the black was obviously the safest option, but it also meant agreeing to be spectators to the macabre show unfolding beneath them. The alternative was for Granite Mountain to follow the ridge out to their escape route and to the safety zone at the Helms' place.

It was an appealing option. The Helms' place would put the hotshots in a better position from which to reengage the fire. With all its defensible space, the ranch was not under threat, and if they could wait out the firestorm there, the crew would be a short walk away from Glen Ilah, where they could help homeowners whose lives were soon to go up in flames. Moving Granite Mountain was also the type of tactical decision that might surprise and impress Abel and the incident commanders. Granite Mountain had been sidelined on the fire's cold heel all day. But when the flames swept through Yarnell, Abel and Cordes would immediately need all the help they could find. For Granite Mountain to emerge unexpectedly into the action, just minutes after the fire had torn through Yarnell, would be a slick move—a coup that could win an ambitious division accolades with the incident management team or a recently absent superintendent the admiration of his unfamiliar crew.

But moving Granite Mountain to the Helms' place came with substantial risks. It compromised many of the Ten and Eighteen that both Marsh and Steed had memorized:

> Weather is getting hotter and drier.
> Wind increases and/or changes direction.

Terrain and fuels make escape to safety zones
    difficult.
Unburned fuel between you and the fire.

The safety zone was more than a mile and a half away from the crew, and as soon as the hotshots left the safety of the black, a sea of dry chaparral would lie between the men and the fire.

Steed and Marsh considered the rules but found ways to justify ignoring the Ten and Eighteen. From the escape route along the ridge, the hotshots had an expansive view of the fire's spread. If the fire got too close, they could always bail off the south, southwest, or west side of the Weavers into the thinner fuels in the desert below.

Ultimately, the choice was Marsh's, and the orders Donut heard him deliver were clear: *Move the crew along the escape route.* If Steed or anybody else questioned his decision, they did so discreetly, because on the crew's radio channel, nobody openly disagreed with Marsh.

From his lookout on Highway 89, Cordes was scrambling to come up with a plan to keep the fire out of Yarnell. He radioed Frisby, who was still in the ATV, racing along the network of dirt roads to pick up his captain, Trew, near St. Joseph's Shrine.

"Is it still an option to burn off that dozer line?" Cordes asked.

Frisby replied that it was not. The wind was too strong, and the flames too close to safely put firefighters before them. From his vantage, Marsh confirmed Frisby's judgment.

"The fire has progressed nearly to the buggies," Marsh said over the radio. Then he came back a moment later. "I want to pass on that we're going to make our way to our escape route and to our pre-designated safety zone."

When Cordes heard the transmission, he wasn't certain where Granite Mountain was, where Marsh was in relation to the hotshots, or even what escape route he intended to take. But upon hearing the phrase "pre-designated safety zone," Cordes immediately assumed that Granite Mountain was going to the Helms' place—the safety zone they'd discussed that morning. If Marsh felt they had the time to get there, Cordes saw no reason to doubt that their escape route was safe and justifiable.

Frisby, though, didn't follow Marsh's intentions. He attempted to clarify.

"Are you in good black?" Frisby asked.

"Picking our way through the black to the road in the bottom, then out towards the ranch," Marsh said. He was breathing hard. Marsh was referencing the two-track that led from the fire's cold heel and back along the top of the long ridgeline to the Helms' place. It was the most direct route back to Granite Mountain's pre-designated safety zone, and the one he'd discussed with the hikers when he had first arrived at the fire's edge that morning. But there were no clues in Marsh's transmission to suggest that this was his intended escape route.

"To confirm, you're talking about the road you saw me on with the Razor this morning?" Frisby asked, confused.

"Yes, the road I saw you on," Marsh said.

This further confused Frisby. It was as if Marsh was being intentionally cryptic. The route he was describing sounded like the road that Frisby had only moments before used to rescue Donut from his lookout. Hiking the crew out that road made no sense whatsoever. The route was, in fact, lethal. It ran straight through the unburned chaparral, and if the brush wasn't already burning, it would be soon.

Regardless, Frisby now had troubles of his own to worry about. Every moment that passed, the fire moved closer to Donut and Granite Mountain's rigs, and he and Trew needed to get them out before the Yarnell Hill Fire claimed its first casualties. Frisby broke off contact with Marsh, trusting the longtime superintendent's judgment. But where Granite Mountain was and where they were heading remained unclear to most firefighters on scene.

Shortly after 4:00, Cordes was still parked on Highway 89 and acting as a lookout for the firefighters in town. The fire had burned northeast to the doorsteps of Peeples Valley before it essentially made a U-turn, and it was now steaming southwest toward Yarnell. Cordes's contingency plan to burn off the dozer line had already failed, and the winds from the outflow boundary were still minutes away from bringing their full force to bear on the fire's northern edge. Once again, a Southwest blaze was defying the expectations of the firefighters on scene. But Yarnell Hill's explosion was well beyond anything Cordes had anticipated.

"Time to get Yarnell on evacuation notice," he called to Abel. "It's at my first trigger point."

Earlier in the day, they'd agreed on four trigger points for the town, each a geographic feature separated by roughly a quarter-mile. The fire had reached the first, a granite knob a mile north of town that meant Yarnell had one hour to evacuate. Already, a reverse 911 phone call had gone out to the homeowners, warning them to get out of town. The next trigger point, a ridge, meant the immediate evacuation of any civilians still left in their homes. The third, another knob, prompted the disengagement of firefighters. And the fourth, a rocky ridge above St. Joseph's Shrine in Harper Canyon, was what Cordes called the "aw shit" point. If flames reached that ridge, every firefighting resource in Yarnell needed to be in or immediately heading toward the Ranch House Restaurant, the predetermined safety zone for the firefighters in Yarnell.

Paul Musser, the second operations chief, was with Cordes on the highway. Musser had already requested that any additional engines in Peeples Valley divert to Yarnell. Though he'd ordered dozens more resources from the SWCC—engines, hotshot crews, water tenders—most were still hours out. Cordes didn't have hours. He needed firefighters immediately. Musser called Division Alpha. He wanted hotshot crews. Granite Mountain and Blue Ridge were the only two on scene.

"Are Granite Mountain and Blue Ridge still committed to the ridge?" Musser asked.

Marsh told him Granite Mountain could not get down to help but that Blue Ridge might be available. At that time, Donut, Frisby, and Trew were parking Granite Mountain's vehicles at St. Joseph's Shrine in Harper Canyon. The gray overcast light now looked like inky dusk, but with a shade of orange. The leaves of the oak trees shading the canyon twitched and rustled. When some of the more veteran hotshots on Blue Ridge exited the buggies, they commented that they could feel the column sucking air upward. A few wondered if they were about to witness another Dude Fire. They were standing in the bottom of what would become another inescapable canyon if the column collapsed.

"Buggies are parked," Donut radioed to Steed. "I'm with Blue Ridge. If you guys need anything, let me know," he said.

"Copy. I'll see you soon," Steed said.

CHAPTER 21

THE SPACE BETWEEN

By midafternoon, the aerial firefight had become so complex that two separate planes, both carrying two- or three-person teams, had divided up Air Attack's responsibilities. Only one plane was in the air at a time, and they rotated out when the aircraft ran out of fuel or the pilots' Federal Aviation Administration–regulated flight time expired. At 4:13, the Air Attack was Bravo 33, which also went by the moniker Arizona 16. The high-powered prop plane was orbiting around the smoke column, and on board were three highly trained aerial firefighters led by a New Mexico–based specialist named John Burfiend. By then, the winds of the outflow boundary had hit the fire's northern edge at Peeples Valley, and the already erratic fire behavior was becoming extreme. The outflow winds would arrive at the fire's southern end in only ten minutes. Burfiend called Marsh to confirm that Granite Mountain was in a good place to weather the coming blow-up.

"Division Alpha, Bravo 33. What's your location?"

Marsh answered, short of breath, "Uh, the guys—uh, Granite—is making their way down the escape route from this morning. It heads south, midslope cut road. We're ma—making our way down into the structures."

It sounded as if Marsh was still on the ridge, out ahead of Granite Mountain, and because he'd made it clear that he and his hotshots weren't threatened, Bravo 33 quickly put the transmission out of mind as he moved on to confirm the safety of the next resource.

Steed and the men were on the escape route, more than a half-mile still from the safety zone. As they contoured southeast along the two-track, the winds blew across Grant's left cheek and toyed with the tails of rolled flagging the hotshots had affixed to their packs. Grant saw the flames, now well past Donut's lookout and the place the buggies had parked. They were rushing toward Yarnell.

As the men hiked down the boulder-strewn two-track in a single-file line, Grant looked down at the boot heels of the man in front of him, trying to avoid rolling his ankle on the rounded stones cluttering the path. Steed and Granite Mountain had been moving for about fifteen minutes when the road hooked to their right above a canyon that faced Yarnell's Glen Ilah subdivision and the Ranch House Restaurant. Granite outcrops and enormous boulders formed the basin's off-vertical walls, and in the canyon bottom, game trails cut through brush that grew thickest near the drainages and swales that flowed toward Yarnell.

The Helms' place sat just beyond the head of the canyon. It was the first time they'd seen it, and the ranch did look big. Multiple barns and animal pens were scattered across the spread, and light glinted off the buildings' metal roofs. At this point, Marsh, Steed, and the crew had to make another decision. They could either stick to the two-track that wound around the ridgeline for another mile before emptying into Glen Ilah, or descend five hundred feet into the basin and go directly to the Helms' place. Cutting through the basin could save the crew time. It looked like a fifteen-minute hike—maybe less.

Seven minutes after the evacuation of Yarnell began, the fire hit Cordes's second trigger point. The homeowners needed to leave immediately. Eight minutes after that, the flames overran Cordes's third trigger point, and he gave the order for all firefighters in Yarnell to disengage. In just fifteen minutes, the fire had covered a mile of ground, a distance Cordes had calculated would take four hours to burn.

Aerial support was one of the firefighters' last good tactics for checking the flames before they reached town, but the smoke was too thick for Cordes or anybody else in Yarnell to help coordinate tanker and helicopter drops from the ground, which meant that Burfiend and Bravo 33 would have to take on the responsibility. It meant more radio traffic for Burfiend, who was already scanning as many as seven different frequencies and witnessing a general degradation of order that he must have felt some responsibility to turn around. He was fielding multiple calls a minute about houses that were soon to burn and air-tanker and helicopter drops.

They were losing Yarnell. Cordes's "aw shit" ridge caught fire as the outflow boundary slammed into Yarnell, and he gave the order for all firefighters to pull out and move back to the safety zone at the Ranch House Restaurant. Donut and the Blue Ridge Hotshots had just started clearing brush around the houses, in a desperate effort to protect a few more structures, when Frisby and Trew told the hotshots to run back to the buggies and get out. Once they were loaded up and heading to the safety zone, Trew and Frisby turned their attention to herding all the remaining firefighters from the canyon. Forest Service engines and a few volunteers from Peeples Valley were also packed into the dry wash.

"Leave! Get out now!" they yelled to the fleeing firefighters.

One volunteer from the Peeples Valley Fire Department refused. He was waiting by his engine in a meadow beneath "aw shit" ridge. Frisby had to yell over the wind.

"I want you to get that truck out of here now," Frisby told the volunteer.

"I can't do that," he yelled back. "There are four men still a mile out in the forest."

The leading gusts of the outflow boundary now hit the canyon. Flames rolled toward the firefighters with such intensity that the volunteer would liken the sensation to being snuffed out by the hand of

God. Smoke blotted out the sunlight, and the canyon filled with flames. Embers rained down around them, and within seconds, spot fires were growing from the size of a quarter to ten-by-ten-foot blazes and spreading rapidly.

"You need to leave right now!" Frisby yelled to the volunteer. If he didn't, he might never be able to. He stopped resisting and fled with his engine to the safety zone. In search of the missing Peeples Valley firefighters, Trew and Frisby took the Razor still farther up Harper Canyon. Flames were consuming brush and trees on both sides of the road when Blue Ridge's overhead found four men hiking down the sandy drainages. *"Go, go, go!"* Frisby and Trew yelled, spurring the men to move faster.

The volunteers ran toward the trucks, near the base of the canyon, with their heads tucked into their shoulders to shield themselves from the heat. They loaded into the trucks, coughing, as the drivers sped out of Harper Canyon through smoke so thick that even with their high beams on, they couldn't see much beyond the hoods.

The Helms' place was always in view at the head of the basin, but the lower the crew dropped down, the farther away their safety zone seemed to be. What had looked like a half-mile from the two-track now looked like three-quarters of a mile. The enormous boulders and head-high brush on the basin's walls had played a trick on the hot-shots' eyes, squeezing the foreground and background into a deceptively shallow plane.

Steed led the men down a draw that at points became so steep, the sawyers and scrape had to balance their saws and tools on the uphill sides of the trunks of the forty-five-year-old scrub oaks. Pebbles and sand strewn across the granite boulders made each step treacherously slick. One at a time, the men lowered themselves down the bedrock steps, using the brush as a makeshift rope.

When the sawyer ahead of Grant moved through one particularly steep section, the crew's new swamper waited his turn in line. Grant looked out at the blooms of black smoke rolling up the edges of the column. He could see the owners of the Helms' place hurrying to push their llama and their miniature donkey into the barns. He could see emergency lights strobing as fire engines raced through Yarnell, and he could see the tankers, helicopters, and Air Attack plane circling above town. The men and women in the planes had no reason to be looking down at the hotshots. The crew had told nobody of their plans to drop into the canyon, and as such, they had received no warning about how far the fire had advanced since leaving the two-track. Grant and the rest of the hotshots couldn't see the flames. The ridge to Granite Mountain's left blocked their view and sheltered the crew from the wind. But in the few minutes since they'd left the two-track and lost sight of the flames, the outflow boundary's winds had moved nearly two miles south of Peeples Valley. Gusts of twenty to thirty miles per hour were now passing over northern Yarnell, and a wall of flames had pivoted, turning toward the Helms' place and the canyon.

The fire's size had nearly doubled since 3 P.M., shooting the column up to thirty-eight thousand feet—nine thousand feet higher than Mount Everest. Up in the stratosphere, the supercharged mixture of heat from the flames, moisture released from vegetation, and turbulent winds had developed into a thunderstorm that now sat perched atop the column. As it had on the Dude Fire, the storm was reaching maturity. Back in town, as Donut drove Marsh's superintendent truck to the safety zone at the Ranch House Restaurant, a barely perceptible rain began to fall, leaving ashy drops on Donut's windshield.

"We've got to go now!" Truman screamed to Lois. She was carrying an armful of things out of their Glen Ilah home to the Winnebago.

"Why?" she asked. Whatever Truman was so panicked about could wait a second. His rudeness had no justification.

"Because if we don't, we'll die!" he yelled.

She glanced out their picture window and saw smoke blowing through the oak trees a few blocks to the north. Suddenly, she understood.

Truman rushed inside, grabbed a plastic bottle filled with holy water, splashed it around the house, and said a quick and silent prayer. He went to shut the garage door. It wouldn't shut—the power had gone out. He didn't have time to deal with it. Outside, a white truck was circling Glen Ilah's streets with a local nurse practitioner leaning out the window, warning people to get out.

"Bravo 33, Division Alpha," Marsh radioed out.

Just after 4:30, the Air Attack plane passed over Marsh and the basin en route to Yarnell. Marsh had seen the plane and knew that retardant along that drop path would help keep flames out of Yarnell.

"That's exactly what we're looking for," Marsh said. "That's where we want the retardant."

Steed and the crew were near the basin's bottom, walking along faint game trails that threaded through the thickest brush. The Helms' place now lay just a third of a mile away. At Steed's normal clip, even in such thick brush, the hotshots could cover that distance in five minutes.

By the time Marsh finished his radio transmission to Bravo 33, the winds of the outflow boundary had passed through the long valley that ran parallel to the ridge Granite Mountain had worked on all day. Rock outcrops and Donut's lookout knob sheared the winds and the fire into a forked tongue. One of the fire's prongs churned up the saddle where Marsh and Steed had met with Blue Ridge that morning. At the ridgetop, two-hundred-foot flames towered above the crest. The fire's second prong raced toward Glen Ilah and the Helms' place. When the winds reached the canyon, the constricting terrain formed a natural chimney, and the gusts accelerated to fifty-two miles per hour.

The gale pulled dry leaves from the chaparral and picked up sand from the basin floor and flung the debris toward Steed, Grant, and the rest of the crew. When the wind first hit, some men turned their shoulders to the gusts. Others slapped a hand to their heads to keep their helmets from blowing off. The high winds were deeply disconcerting, but not yet reason to panic. Steed and the hotshots had seen the fire before they dropped off the ridge. It was almost a mile away and moving southeast into Yarnell—not toward the basin.

But moments later, smoke blew from left to right across the box canyon's exit, cloaking the hotshots' view of the Helms' place. Still, Steed kept hiking. It could be drift smoke. *Let it be drift smoke.* Darkness fell over the canyon. At the first flash of orange, Steed knew what was coming.

"Breaking in on Arizona 16, Granite Mountain Hotshots!" he shouted into his radio, his voice quaking with emotion. "We are in front of the flaming front."

Arizona 16—John Burfiend, also known as Bravo 33—didn't respond. At the moment of Steed's radio call, Burfiend was in the process of guiding an expensive DC-10 into a retardant drop designed to save a few hundred homes. But Steed's overmodulated and barely comprehensible mayday call prompted Burfiend to call off the slurry drop. He didn't know where Granite Mountain was or the extent of the crew's problem, but he recognized panic when he heard it, and he knew that the ten thousand gallons of retardant in the plane's belly might prove critical.

It took Burfiend seven agonizing seconds to send the DC-10 back into an orbit overhead and switch radio frequencies to respond to Steed. That seemed too long to Abel. He'd heard the urgency of Steed's transmission. Abel stepped in to help.

"Bravo 33, Operations. You copying that on Air to Ground?"

Air to Ground 16 was the radio frequency Steed had used to call to Burfiend. Abel shortened it to "Air to Ground" and referenced the frequency to help Burfiend decipher, among the absolute audible chaos emitting from his radio, which channel he needed to pay closest attention to. At that moment, all seven of the channels he was scanning seemed to be emitting an unbroken stream of pressing radio transmissions from both the nine different aircraft he was in charge of and the ground resources who needed their support. Before Burfiend could respond to Abel, Marsh broke through the cacophony with a call from the basin.

"Air to Ground 16, Granite Mountain. How do you read?!"

Marsh's moniker was actually Division Alpha, but at the sight of a flaming wall rushing toward him, he defaulted to the one he was most familiar with—Granite Mountain. Firefighters on scene immediately recognized Marsh's southern drawl. Talking among themselves, one asked, "Is Granite Mountain still out in the green?" Another who'd paid close attention to the radio throughout the day knew the answer. Division Alpha had clearly said that Granite Mountain was in its safety zone: the black. What Marsh hadn't been clear about was the fact that the crew was leaving the black for the Helms' place.

In the nine seconds since Steed had made his mayday radio transmission, a sixty-foot wall of flames had swept across the canyon's mouth and advanced some 150 feet closer to the men. At a glance, the crew comprehended the horrors about to unfold. Fire blocked their escape from the basin. The wind seemed to be throwing the flames, and the fire leaped between bushes as it rushed toward the hotshots with the fluid intensity of a breached dam. The hotshots could not outrun the flames. The fire was spreading at ten to twelve miles per hour, and the wall they'd just hiked down—the only escape not blocked by flames—was too steep to run up. Their only chance for survival lay in deploying their fire shelters.

Steed ordered the men to cut a deployment site at a slight depression where the vegetation grew thinnest, near two swales that drained out toward Yarnell. Scott, Andrew, and Wade ripped their chainsaws to life and fanned out into the chaparral surrounding the depression. Their saws screamed, and Grant and the swampers followed behind, frantically tossing brush into wind so strong it swept the cut bushes clear of the deployment site before they could hit the ground. The scrape raked the flammable leaf litter off the deployment site, and every glance down the basin affirmed dark fears. The flames grew to eighty feet long and tilted nearly horizontal to the ground. Debilitating blasts of heat rushed uphill. Behind the hotshots, a deer tried to scramble up the canyon walls.

"Air Attack, Granite Mountain 7." This time it was Bob calling

Bravo 33, using a stress-driven amalgamation of Crew 7 and Granite Mountain. "How do you copy me?!"

Firefighters in Yarnell heard in the background of Bob's transmission the sinister screaming of chainsaws, which a few recognized as a sign that the crew was cutting a deployment site.

When Burfiend finally responded to the calls, he followed the chain of command. His reply to Todd Abel carried the clinical tone a 911 dispatcher adopts to calm down a panicked caller. It came across as long-winded.

"Okay, I was copying a little bit of that, uh, conversation, uh, on Air to Ground," Burfiend said. "We're—we'll do the best we can. We got the Type 1 helicopters ordered back in. Uh . . . we'll see what we can do."

"Bravo 33, Operations on Air to Ground." Abel was getting irritated now. He wanted Air Attack to address Granite Mountain.

No response for six more seconds. Bravo 33 was still fielding other radio traffic. Finally Burfiend responded to Abel: "Operations, Bravo 33."

Now Bob broke in, his voice betraying intense frustration. "Air Attack, Granite Mountain 7!"

A minute and a half after Steed made the mayday radio transmission, Burfiend responded to Granite Mountain. The flames were within a couple hundred feet of the men now.

"Okay, unit that's hollerin' in the radio," Burfiend said, "I need you to *quit*." Then he took Abel's call.

"Granite Mountain 7 sounds like they got some trouble," Abel said, and Burfiend immediately responded that he'd "get with Granite Mountain 7 then."

But Burfiend never talked to Steed or Bob. As the flames rushed closer, it was Marsh who finally contacted him.

"Bravo three th*rrreee,*" Marsh said, drawing out the last word. He sounded completely composed at first. "Division Alpha with Granite Mountain."

"Okay, Division Alpha, Bravo 33."

"I'm here with Granite Mountain Hotshots," Marsh said. His

voice grew shaky. "Our escape route has been cut off. We are preparing a deployment site. We are burning out around ourselves in the brush. And I'll give you a call when we're under the—the shelters."

"Okay, copy that. So you're on the south side of the fire then?" His tone changed. Burfiend understood now.

"Affirm!"

Steed yelled the order to deploy against winds howling so loud, they stole all other sounds. Embers, riding debilitating blasts of heat, struck the men with the velocity of quarters thrown from a speeding car's window. Out of time, Scott and the other sawyers pulled back into the deployment site they were able to clear in the three minutes they had. It was half the size of a tennis court. As they ran back toward the site, most of the hotshots threw their chainsaws, tools, and packs, but Scott kept wearing his. As he sprinted into the tiny clearing, he unclipped his waist belt, dropped the pack, and pulled the shelter from its red-cloth casing.

Steed was just in front of Scott on the far right edge of the deployment site. To his left was Joe Thurston, Zup was behind him, and Bob was to his right. All of them were wrestling with their shelters, which thrashed into one another like wind socks in a hurricane-force gale. Scott put his back to the fire, grabbed the shelter's handles, and flicked out the aluminum tent. The wind tried to rip the shelter from Scott's hands. He pulled the metal sack over his head and fell face-first into the granite cobbles. Inside the tent, he had shelter from the wind. In the scramble, a small yellow-handled hand tool one of the swampers had carried in his pack had dropped to the ground, and Scott fell on top of it. He put his feet toward the flames and, with his elbows and boots, pinned down the aluminum flaps on the shelter's underside. He held tight to the handles on the inside and cursed himself for taking his leather gloves off earlier in the day.

The flames rushed up the basin, only feet away now. Inside the shelters, it glowed orange, and the superheated wind slapped the aluminum tent against Scott's helmet and back. In his group, the men now lay so close that Scott's feet brushed against Joe Thurston's, his right shoulder nearly touched Bob's left, and his head was just a foot from Steed's hip. They huddled there together, nostrils filled with dust and heat, and maybe yelled encouragement to one another. *Hold on! No matter how bad it hurts, stay in your shelter! We got this!* Or maybe they simply clenched their jaws.

Grant was almost six feet uphill from Scott's group. He was alone and deploying farther from the flames than any of the others. He dropped his pack beside himself, tugged the shelter from its sleeve, and watched in despair as it snapped out into the wind. He struggled to pull the heat shield back to his body. The flames roared closer. Grant turned away from the fire, and again the shelter leaped from his hands. He pulled it close again, and this time slipped his left foot inside the aluminum sack. By then the flames were already lashing at the crew. The fire first hit Marsh, who had deployed closest to the canyon's head, and in seconds it rolled over Scott's group before finally consuming the brush surrounding Grant.

CHAPTER 22

CROSSING THE LINE

Over the next six minutes, Burfiend made seven attempts to reach Division Alpha and Granite Mountain 7 while he simultaneously tried to guide a helicopter to drop water on their site. But he didn't know precisely where the men were. The deployment's effect on the firefight was immediate; saving Granite Mountain became the sole focus. At 4:40 on June 30, some 340 firefighters were battling Yarnell Hill, and the radio traffic was unrelenting. Cell phones became the firefighters' most reliable method of communication. Everybody wanted to reach somebody who could tell them something about what was happening. But the only thing anybody knew for certain was that Granite Mountain had deployed during the most maniacal fire most of the firefighters had ever seen.

Darrell Willis was still up in Peeples Valley when Todd Abel called him and gave him the news. Willis handed off his structure-protection duties and headed straight for Yarnell.

The scene in town made it clear just how unhinged things had become. A pair of panicked horses galloped down the middle of Highway 89 while a shocked engine company stood on the edge of the road. The fire, of course, hadn't stopped advancing after Marsh's last radio communication. Dozens of homes were on fire, and the flames reached nearly to the antiques stores in downtown Yarnell. A helicopter tried flying straight through the column to locate the men on the opposite side of town, and the DC-10 flew in a holding pattern overhead. The hope was that 10,743 gallons of retardant dropped directly on the crew could save their lives, but with fifty-mile-per-hour winds coursing through the canyon, it would have been nearly impossible for a DC-10 to drop accurately. Even if the pilot hit his target, the gale would likely blow the liquid away before the slurry landed on the shelters.

After the burnover, the Ranch House Restaurant became the Incident Command Post for managing the rescue. The safety zone was a somber scene amid the chaos. Across the street, a long line of cars, the Ferrells among them, exited Glen Ilah. Each driver paused dutifully at the stop sign while in their rearview mirrors they could see houses engulfed in flames and the fire steadily approaching. Granite Mountain's and Blue Ridge's buggies, along with almost thirty other fire vehicles, were parked in the restaurant's lot.

Abel and Musser were in their trucks looking out at Glen Ilah. Flames and smoke blocked their view of the long ridgeline of the Weavers. Somewhere out there, they knew, Granite Mountain lay burned beneath their shelters, no doubt with critical injuries and very possibly worse.

"What's going on?" Willis asked when he pulled in.

"We can't get ahold of them on the radio," Abel said. He switched frequencies on his truck radio and tried again. "Granite Mountain, Operations." Every few minutes, he repeated the process. Trew, the Blue Ridge captain, sat nearby in one of Granite Mountain's trucks, trying to hail the men on the crew's personal radio network. Abel heard nothing back. The only response Trew got was the pulsing static of a keyed mic.

Abel and Musser pulled out a procedural guide for managing an incident within an incident. They needed to keep fighting fire while organizing a rescue mission that required first responders to head directly into a firestorm. They'd already divided up the responsibilities. Musser would run the firefight while Abel managed the rescue under the title Granite Mountain IC. He'd already notified the Phoenix burn center and had five medevac helicopters, plus ambulances, en route to transport survivors. At that point, Abel had three goals: Find the men, get them to a hospital, and prevent anybody else from getting injured in the process.

"You got a crew member over there with Blue Ridge," Abel said to Willis, nodding toward Granite Mountain's buggies.

A few of Blue Ridge's guys sat outside Marsh's superintendent truck, toying with the loose stones at their feet. Donut was with them. The Blue Ridge hotshots had turned their radios off to shelter him from any more news, but Donut had already heard the deployment traffic en route from Harper Canyon to the Ranch House Restaurant.

Donut's first thought was *Why am I here?* Followed closely by *There's no way they could survive that.* His immediate duties provided some distraction. He kept driving, following the glowing brake lights of the Granite Mountain buggy in front of him. Once they'd pulled into the Ranch House, the Blue Ridge guys asked him to get the medical kits and oxygen bottles off Granite Mountain's rigs. But that flurry of activity was short-lived, and, in the absence of a job, Donut was left with time that refused to pass.

A medical team assembled near Donut's truck. He could hear a supervisor addressing a group of EMTs.

"If you do not feel comfortable going in, I need to know right now," the supervisor said. A propane tank vented by one of the burning houses, and a spindle of flame, whistling like an exponentially more powerful bottle rocket, shot thirty feet into the air.

Willis hugged Donut.

"How you doing?" he asked.

"All right," Donut said.

A long moment passed, but neither said anything. Donut was in shock. Willis wasn't far from it. He squeezed Donut's arm and walked away to call Prescott's fire chief. The uncomfortable truth was that the City of Prescott needed to ready itself for a coming storm of a completely different nature.

Meanwhile, the flaming front slammed into Glen Ilah. A few houses down from where the Ferrells lived, Bob Hart, ninety-four, thought little of the fire. He hadn't received the reverse 911 calls that went out that afternoon warning the citizens to evacuate, no fire or police personnel or neighbors had stopped by to tell him of the coming storm, and though it's possible he'd seen the fire glowing on the hillside at some point during the three days it had been burning, Bob, unlike the Ferrells, never considered it a real threat. That is, until he went to the kitchen and saw through the window flames consuming trees and homes. He called to his eighty-nine-year-old wife, Ruth, and they rushed out of the house, but the garage door opener didn't work. *Click, click, click*: Nothing. Together, Ruth and Bob managed to push open the garage door and they climbed into the truck.

Bob couldn't see through the smoke. He kept bumping into the trees and brush on the sides of their driveway. Then he put the truck's right wheel into a ditch. The tire exploded. Around them, dozens of propane tanks sent columns of flame shooting into the air like fires off an oil derrick. The Harts could drive no farther, but because the truck's engine was still running and the air-conditioning kept working, they opted to stay inside the truck. For a moment, it seemed like the best option. But when the flaming front swept through the Harts' backyard, Ruth, still not more than a few hundred yards from their home, decided to walk out of the situation or die trying. She refused to burn while sitting passively in the truck.

Cordes, the structure-protection specialist, was driving through Glen Ilah in search of civilians. When he found the Harts, they were walking hand in hand through the smoke, wearing only pajamas.

"Could we please get a ride?" Ruth asked him. Cordes took the couple back to the Ranch House, where they sat on a bench for the next hour and a half. Cordes turned around and drove back to Glen Ilah.

He'd first entered the subdivision shortly after hearing about Granite Mountain's deployment. While Abel dealt with the logistics of the rescue, and most other firefighters waited with dread for news about the hotshots, Cordes went back to Glen Ilah to help civilians still in the subdivision.

He took a different road on his second trip. People were everywhere. Cordes pulled up alongside them and yelled from his window what seemed plain: "Leave the area immediately!"

Many simply refused, their reasons bizarre. One man pushing a goat into his car told Cordes that he wouldn't escape until the animal got inside. Another man ran back inside to fix a sandwich. A sixty-year-old woman and her thirty-eight-year-old daughter told Cordes that their neighbor had thirty-five cats and dogs locked in the house, and if the pets were going to burn, so would they.

"You *are* going to die if you don't exit the area immediately," Cordes told them, exasperated. "I cannot emphasize that enough."

Cordes made a final plea, which they refused, and he headed deeper into Glen Ilah. He was nearly to the place where he'd shown Marsh the safety zone that morning when a man waved him down. When he opened the window, smoke burned Cordes's eyes.

"I can't leave the area," the man said, yelling over the wind. "My neighbor's still in there. She's handicapped. I can't carry her any farther."

The storm's full force was now upon Glen Ilah. Forty-mile-an-hour winds hammered the subdivision. At points, the flames curled over the road, and Cordes drove on below them. Through a window in the blowing smoke he saw in his high beams the old woman lying on the ground. Cordes got out of the truck and picked her up. He carried her along the side of the truck to shelter her from the wind and, with one hand, opened the door and set her inside. Meanwhile, the man he'd just talked to opened the opposite door.

"Shut the fucking door!" Cordes yelled. Like a vacuum, the calm air in the cab sucked embers into the truck, and they landed on the old woman's chest, face, and lap. She screamed, but Cordes couldn't help with the pain. The fire was still pushing toward them. He slammed the door shut.

"Get in!" Cordes said to the man. Already blisters boiled where the embers had struck his face. Cordes raced out the roads, past lingering civilians, to the Ranch House, where the burn victims were loaded into an ambulance and taken to the hospital. Once again, Cordes tried to head back into Glen Ilah, but he couldn't. Flames had jumped Highway 89 and surrounded the Ranch House. By night's end, whatever the fate of Granite Mountain, Cordes was certain he'd be pulling bodies from Glen Ilah.

Trew and Frisby were at the Ranch House with three other firefighters from the Prescott National Forest. They had three ATVs and planned to drive into the blaze to find the hotshots. For the fourth time in thirty minutes, the fire bumped into Highway 89, and an engine crew was spraying foam on the restaurant.

At the first sign of calm, Trew, Frisby, and the men on the other two ATVs headed back into Harper Canyon, where St. Joseph's Shrine and its marble statues of Jesus were blackened but still standing. Winds whipped the tallest trees, and the fire had burned a few houses nearly to the ground—their chimneys now rose above piles of smoldering ruins. Metal wind chimes clacked in the front yards of torched homes. Trew and Frisby stopped at a wide point in the road with fire on all sides and looked across at the dozer line that led out toward Donut's lookout and the saddle beyond. It was too hot to proceed.

For six minutes, they waited in the pseudo safety zone. The fire was slowly calming down, and the road formed a keyhole through the flames, which were no more than a hundred yards deep. Just beyond the venting propane tanks and smoldering brush lay the cooler black, and their only chance at finding Granite Mountain soon.

"Fuck it," Trew said finally. "Let's go."

He stomped on the gas, and they tore up the dozer line. Burning trees had fallen across the road, and Trew and Frisby ducked under low-hanging power lines. Not far beyond Harper Canyon, the road opened up into the long valley. It was a moonscape. Small braids of white smoke rose from a sea of black that stretched from ridge to ridge. Donut's lookout, the grader's aged metal, the slopes where

Steed and the crew had been cutting line—all of it was blackened. What remained were hot embers glowing in pits of heat and the twisted and grotesque spent matchsticks of the oldest oaks and chaparral. The three ATVs entered the great emptiness in a caravan.

Back at the Ranch House, Donut called his mom. He didn't know what else to do. He told her something had happened. That it was bad and it could get worse but that he was okay.

He couldn't talk yet about what had happened. He simply wasn't capable of comprehending it himself. But even if there was nothing to say, just having his mom on the phone was grounding. Donut was in the front seat of Alpha's buggy when a firefighter he didn't know asked him for the crew's manifest, the names and weights of everybody on Granite Mountain. He handed them a list, which had been compiled during the Doce Fire, from the clipboard in the center console.

Eric Tarr was the medic in the front right seat of Ranger 58, a helicopter dispatched from Peeples Valley to locate the hotshots. Tarr had spotted yellow packs on the ridgetop near the saddle where Marsh had met with Trew and Frisby, who were now racing toward the ridge on their ATVs. He radioed to them the coordinates of the packs, and when the road became too rough to proceed, the five men jumped from the vehicles, grabbed a backboard and medical kit, and ran three hundred yards up toward the yellow bags. But this didn't seem to Tarr like the right place.

The site didn't fit with what he'd heard Granite Mountain describing on the radio before they'd deployed: that they were on a two-track that ran to the south, that some safety zone was a half-mile out, that structures were nearby. None of the transmissions had been particularly clear, but they pointed to somewhere other than the ridgetop.

It was just after 6 P.M., almost an hour and a half after Granite Mountain had deployed. In the distance, flames were still working through homes in Yarnell. Up until then, the smoke to the southwest had been too thick to fly through, but as the winds ceased, the smoke began to disperse, and in a fleeting window Tarr saw the Helms' place appear. He nudged the pilot to fly south along the two-track, and where the thin road hooked to the right and the canyon fell away beneath the helicopter, Tarr saw the glint of aluminum: the shelters.

He radioed the coordinates to Air Attack. The spot where the helicopter managed to land was four to five hundred yards from the shelters. Eddies of swirling ash were kicked up by the rotors as Tarr exited the helicopter.

He carried with him only his fire shelter and a small medical bag. Each footstep broke through the crust of the blackened soil, which had a texture like frozen-over snow. The pilot, running perilously low on fuel, returned to base, and Tarr was left alone in the basin.

The world existed only in absolutes of black and white. Faint white smoke hung in the air. Like the shells on boiled eggs, the fire's extreme heat had cracked the massive granite boulders scattered about the basin floor. Tarr breathed through his CamelBak tube, using the water stored in the plastic bladder to cool the remnant heat of the basin.

He passed a chainsaw, the plastic gone, and a Pulaski head but no handle. He saw the melted remnants of a pack. Frayed fabric encased batteries and files, and the chains and wire-mesh goggles once stored inside. He saw exploded fuel bottles that had been tossed aside just moments before the men deployed, and, twenty yards from them, he saw the site.

The still-intact shelters were nothing but pale silica cloth sacks completely stripped of their aluminum. Tarr yelled.

Hello! My name is Eric Tarr. I am a medic. I'm here to help.

He couldn't yet bring himself to go into the site. He was either alone in a box canyon with nineteen dead men or alone in a box canyon with a few critically burned survivors. At first, the only response to his calls was the echo from the granite walls, but then he heard a quiet click and a perplexing murmur of voices. Tarr moved closer.

The shelters told the story of rushed but organized deployment. Nobody had tried to flee the deployment. To Tarr's right, one group of five hotshots still lay beneath their intact shelters. But most of Granite Mountain's men lay in the open.

The first body Tarr came upon had deployed closest to the flames. The man lay facedown outside of his shelter, which lay contorted and twisted along his left side. Beside him was a puddle of red plastic. It had been Marsh's helmet. Tarr checked for a pulse and found none. He stepped farther into the deployment site.

Five more firefighters lay in similar or worse condition to his left. Tarr would later note that "each of them appeared similar with obvious rigor and no breathing or signs of life."

But the crew member with the most severe burns lay the farthest from the flames. His body was faceup on the remnants of his pack. All that remained of his clothing was a small triangle of yellow Nomex on his back, a strip of his leather belt, and the rear pocket of his pants. Because he checked all the men's pulses, Tarr checked this man's, but he didn't need to.

If there were survivors, they'd be in the group of five tightly deployed shelters that were still partially intact. Somewhere from this group had emanated the strange clicking noise followed by a murmur of voices. It was the hotshots' radios. They were still functioning. Tarr couldn't see the firefighters beneath the shelters. He spoke again.

Hello. I'm Eric. I'm here to help. Hello.

No response. On the two-track above the basin, four firefighters appeared. One was Trew. He was now scrambling down the hill toward the site. Tarr bent down and raised the intact cloth. The man inside lay on his stomach within a perfectly deployed shelter that had been all but destroyed. His hands were severely burned, but his clothes were intact, and below the body was the yellow plastic handle of a tool. Tarr gently rolled him over. He placed his finger on the wrist. The fire had been far too hot.

Tarr again counted the bodies and stepped reverently outside the circle.

"I have nineteen confirmed fatalities," he radioed out. It was 6:35 P.M.

CHAPTER 23

AFTER

Word of the Granite Mountain Hotshots' deployment was spreading fastest via social media. Already, some of the families had heard rumors of firefighters injured on Yarnell Hill. Up in the basin, a phone buzzed underneath Scott's shelter. His mom was texting.

"Many people praying for you and crew. We heard you were trapped. We love you."

At 7:30 P.M., a cell phone in the center console of Alpha's buggy vibrated. Donut watched it shake in the cupholder. Clayton had sat in Alpha's passenger seat that morning. Then the phones that had been left in the back of the buggy started. Vibrations, rings, jingles—Donut couldn't take it. He stepped outside.

Upon Tarr's confirmation of nineteen fatalities, Donut had gone from a seasonal hotshot on a crew of twenty—in a field of fifty-six thousand wildland firefighters—to the lone survivor of a tragedy that was fast becoming a national story. Just hours after the deployment, the Prescott Fire Department's public information officer was getting calls from as far away as Ireland and New York to confirm what Facebook was telling reporters and the world: nineteen confirmed dead. "The one guy they found is doing fine," a local reporter tweeted.

The media coverage was just the start. Not since the 1930s had so many wildland firefighters been killed in a single incident. Even more striking, nineteen was the greatest number of firefighter deaths since 9/11, when 343 perished in the collapse of the World Trade Center and, if it wasn't already, "firefighter" became synonymous with "hero." Even if most Americans didn't know exactly who hotshots were or what they did, the country was highly sensitized to the fate of its first responders.

While Yarnell was still burning, every national media agency in the country—CNN, *The New York Times,* Fox News, *The Wall Street Journal*—was making arrangements to send reporters to the tiny gold-mining town in the Arizona foothills. Donut was the one source every reporter assumed knew what had happened to his crew and why. He was the story.

Brendan "Donut" McDonough. A tall-tale-spinning, hard-partying third-year hotshot with the crew's longest rap sheet and foulest mouth. A twenty-two-year-old with a high school education, a small arsenal of firearms, an outsize pickup, and a rocky relationship with the mother of his young daughter. A recent convert to Christianity. A young man subject to a blindingly complex mix of emotions.

Of course, Donut didn't know everything that had happened to his crew. He couldn't. At that moment, all he knew for certain was that nineteen of his closest friends had been burned to death and the only thing separating him from the same nightmarish fate was Marsh's off-the-cuff decision to make him the lookout rather than any of the other qualified hotshots on the crew. By this tragedy, Donut was made famous: *the one* who had lived when nineteen died. June 30, 2013, Yarnell, Arizona. The fire would define his life as much as theirs.

That twisted reality was still ahead of him. Donut had a pressing issue to deal with first. The manifest he'd given to some firefighter hours earlier had only eighteen names on it. The number didn't agree with the nineteen firefighters confirmed dead in the bottom of the basin. Not long after hearing the news, another firefighter came to the buggy with the list.

Who was on the crew today? Who was with Granite Mountain? they asked Donut.

Donut couldn't process anything. His thoughts kept returning to the phones. Rings. Senseless jingles. Vibrations.

The firefighter asked Donut again, "We need you to help identify the names of your crew members."

Donut went to the back of the Alpha buggy and pointed to each hotshot's seat: Clayton, Chris, Tony, Wade, Garret, Grant, Sean. *Seven men total.*

Then Donut went to Bravo: Bob, Turby, Joe, Scott, Dustin, Woyjeck, Warneke. *Another seven.*

Then the supe truck—Marsh, Ashcraft. *Two.* And the chase truck—Steed, Carter. *Two more.*

Eighteen. The numbers still didn't add up. Donut's math was wrong. Somebody was missing. Then he remembered John Percin. He'd only started fighting fire with the crew two days earlier. Percin wasn't on the manifest. He didn't have a seat in the buggy yet.

Heather Kennedy was home alone, making dinner. At about 6:00, she got a text from her best friend.

"Is your man okay?"

Then another: "Have you heard from him?"

Heather replied, "Yeah, he's on a fire in Yarnell. They're hot and tired." . . . "I talked to him earlier today." . . .

"Why did something happen?"

Jessica: "No, just making sure. I care about him."

Heather: "Aw, thanks. He's good. Getting lots of hours. I feel bad cuz I didn't let him sleep last night. LOL! He's prolly really tired."

At 8:00, Heather texted her friend back.

"OK, so 22 firefighters injured in Yarnell Hill Fire. Fuck."

And then: "TV said they're dead."

Heather was a cop. She'd stood on people's doorsteps many times before and told them their world was destroyed—that somebody they loved was gone. Police call it notifications. It's the force's worst job. A knock came on Heather's door.

The dogs barked. Heather pushed them into the bedroom. The next thing she would remember was that she was lying on the ground with three colleagues standing over her. Then she was on the couch.

"The whole crew?" she asked.

They nodded their heads. She wept but had to keep moving.

"What do you wear to tell your loved one's family that he's dead?" she asked. Heather wore jeans.

She went to his sister's first. She was living with her husband in a house that Scott had bought with money he earned hotshotting. It was three minutes away. Heather locked up the tears, hid the emotion the same way she did at work, and knocked on the door. Warrior mode, she called it.

His sister answered with a delighted smile.

"Scotty's gone, Jo," Heather said.

It took a moment to comprehend. "What?"

"Scottie's gone."

His sister fell. Heather lifted her up, and together they walked toward the living room. Scott's three-week-old nephew, born while he was on Thompson Ridge, lay on the couch. Jo stumbled toward him.

Heather jumped over the coffee table. She pulled the baby away as Jo collapsed onto the couch. Heather watched the baby, his small hands grasping at empty air.

This can't be real, she thought.

Brandon Bunch had watched the column rise over the Bradshaws most of the afternoon. He'd noticed the wind shift when a small American flag in his backyard that had been blowing in one direction turned and blew in the other. But he didn't think much of it.

He and his family had just gotten back that morning from a weekend trip to Lake Mead. It was Jacob's third birthday, and they'd spent the afternoon celebrating in the backyard. Bunch's dad and family had come over. Between the kids on the swing set and meat on the grill, leaving the hotshots had never felt like a better decision.

The family had just left. Bunch was out front watering the plants when his pastor from the Heights Church called, sobbing. The pastor was a close friend of Clayton's and had been asked by the fire department to counsel the families. "Clay didn't make it," he told Bunch. It was all the pastor knew for certain.

Janae stayed at home with the kids, and Bunch left immediately. Families, friends, and fire personnel were meeting at Mile High Middle School. He met Renan and Phillip Maldonado, the squad boss who had quit that spring, and they went over together.

Mile High is a thirties-era brick building that sits a block from Whiskey Row on Granite Creek. The air was damp, and the creek was flowing for the first time that year when the three former hotshots pulled in.

Most people milled around outside by the row of ancient junipers and a wooden carving of a badger pup, the school mascot. Some people were inside, sitting in silence in the school's band room, where children's clarinets and tubas were stuffed into cubbies beneath a poster of Miles Davis. A few, like Bunch, had heard rumors that their loved ones were dead, but the only news the City of Prescott had made official was that Granite Mountain had deployed.

It was dark by the time the city decided that enough people had arrived to break the news. The auditorium's entryway and the nine hundred tattered yellow cloth seats were nearly full, and Kleenex and cases of bottled water had been placed at the doors. A police officer, followed by a small man named Dan Bates, the vice president of the United Yavapai Firefighters Association, walked down the long aisle between the chairs. Few people cried. Most sat wide-eyed and scared.

Bates took his place on the stage between a pair of oversize vases stuffed with plastic flowers. He didn't explain what had happened, because he didn't know. The analog clock behind him read 8:45 P.M., the red second hand sweeping continuously.

"I'm going to read off the names of the deceased," Bates said. He didn't mince words. "Nineteen are gone. Brendan McDonough is the only survivor."

The tears and wailing began at the reading of the first name and rose again at the next. Bates didn't pause.

ERIC MARSH

JESSE STEED

CLAYTON WHITTED

ROBERT CALDWELL

TRAVIS CARTER

CHRISTOPHER MACKENZIE

TRAVIS TURBYFILL

ANDREW ASHCRAFT

JOE THURSTON

WADE PARKER

ANTHONY ROSE

GARRET ZUPPIGER

SCOTT NORRIS

DUSTIN DEFORD

WILLIAM WARNEKE

KEVIN WOYJECK

JOHN PERCIN, JR.

SEAN MISNER

GRANT MCKEE

"Guess we've been here before," Tony Sciacca, the safety officer on Yarnell Hill and the Type 1 incident commander on the Doce, said to Steve Emery. Emery had worked under Sciacca when the incident commander was in charge of the Prescott Hotshots in 1990. They'd worked together on the Dude Fire in Emery's rookie year.

What Emery saw on the Dude Fire triggered what became a decades-long battle with post-traumatic stress disorder. After hot-shotting, he joined the Central Yavapai Fire District, and as a para-medic he continued to see trauma that he couldn't remove from his mind's eye. He turned to alcohol to cope, and in his darkest moments had thoughts of death. Eventually, Emery sought help and learned to live with PTSD, even helping organize a program to help other firefighters deal with the disorder.

Emery was now the second-in-command of a Central Yavapai Fire District engine that had come to Yarnell Hill that morning to help with structure protection in Yarnell. Abel and the incident management team no longer needed firefighters. They needed paramedics.

Emery volunteered for the recovery immediately. He knew the horrors he'd find in the field, but he barely stopped to consider the consequences. He'd deal with the memories later. The men needed his help. After Tarr confirmed the fatalities, Emery called his wife. "The Granite Mountain Hotshots, they're all dead," he told her.

She wept. Her tears were for their marriage as much as the men. She couldn't weather another resurgence of PTSD—it had nearly destroyed their marriage, and the thought of watching her husband suffer again was too much.

"We're dealing with this up front this time," he told her. "We're going to do this right. But I have to go."

Then Emery joined Abel, Willis, and a small clutch of other local firefighters and headed into the basin. Unable to remove the bodies that night, Emery and nine other men spent the night at the Helms' place.

At 1:30 A.M., a dust storm born south of Phoenix passed over the area, spinning small dust devils of ash through the black, while a bulldozer worked through the night to build a hasty road from the Helms' place to the deployment site.

Except for Tarr's cursory evaluation, the men's bodies remained undisturbed. An official investigation needed to take place before they could be removed. At first light, the Yavapai County sheriff and a small forensics team arrived. They mapped the location of the bodies and photographed their equipment and personal effects: Chris MacKenzie's disfigured radio, a Swiss Army knife, a keychain. The bodies were placed inside orange bags and laid together in three rows.

Meanwhile, five more firefighters met Emery, Abel, and the others at the Helms' place. Among them was Wade Parker's father, Danny, a firefighter with the Chino Valley Fire District. He'd asked to be there when his son was transported off the hill.

There was no protocol for retrieving nineteen dead from the side of a mountain. One of the firefighters had been to the World Trade Center after the towers collapsed, so the small group that had volunteered for the grim task arranged a respectful and ceremonious retrieval based on the procedure used in New York. They drove to the site with three pickups. The nineteen lay in orange body bags that were numbered but not named.

The retrieval team took a moment at the site. Some stood over each bag and said a few words, while others walked the basin, aghast. Danny Parker wept. In the long history of fire-line deaths, there had never been a worse place to be than that basin outside of Yarnell.

The sheriff had brought nineteen American flags for the men, and each flag was ceremoniously draped over one of the bodies. Four men, two on each side, lifted a fallen firefighter and cautiously passed the body to three others who stood at attention in the truck's beds. Two fallen hotshots were placed in the back of each truck.

As they returned to the Helms' place, one man rode along in the back of each pickup. He sat on the pickup's rails, his hand placed gently on the bodies to keep the men still and the flags from fluttering as the truck bounced slowly down the track to the safety zone Granite Mountain had never reached.

Removing all nineteen took three separate convoys, plus one more with a single truck. At the Helms' place, Emery and others received the bodies. Then a pastor blessed each hotshot. The careful handling of the men was repeated as they were transferred to medical-examiner vans and driven ninety miles to Phoenix, where the bodies were to be autopsied and prepared for burial.

The caravan of vans and accompanying fire personnel stretched for miles. News of Granite Mountain's deaths had spread throughout the country, and at the base of Yarnell Hill, people began appearing to watch the procession. At first, there were only a few standing along the highway, in a gesture of silent respect. But the crowds grew as the caravan approached Phoenix, where it was 112 degrees. Police cars and fire engines were parked at intersections and overpasses, their officers and firefighters standing in silent salute as the fallen hotshots passed below. Thousands of strangers stood side by side, offering prayers and thanks to the nineteen young men who had unwillingly become heroes.

Donut went to fifteen funerals. On July 10, he went to four: Eric Marsh's was held at Granite Basin Lake at 8 A.M., Wade Parker's at 11 A.M. in Chino Valley, Andrew Ashcraft's at 1 P.M. in Prescott Valley, and Clayton Whitted's at 2 P.M. at the Heights Church. Jesse Steed's was the next day at noon. Chris MacKenzie's was two days later in his hometown in California.

Rain started falling on Prescott the day the hotshots died, and since then, the town had remained wet and brooding. Southwest fire season all but ended on July 1. The monsoon was the heaviest the Southwest had seen in many years. Most afternoons, it gave the town an appropriately damp mood. Shop owners raised banners. Soap was taken to store windows: THANK YOU FIREFIGHTERS . . . OUR HEARTS ARE WITH YOU . . . NEVER FORGET. Granite Mountain bumper stickers appeared on the backs of cars all around the country. The City of Prescott made T-shirts in remembrance of the men, and a loop of purple ribbon with flames in the eye came to symbolize the town's tragedy. So too did the number 19.

Donut said his goodbyes when the men were brought back from the morgue. It was Tuesday morning, two days after the deaths, and Donut went to Phoenix with the department to bring their bodies back to Prescott. This time the procession had a more ceremonial feel.

The hotshots were loaded into nineteen white hearses and lined out in an unbroken single file. Marsh's hearse led the way. Steed's followed. Granite Mountain returned home to Prescott along the same route they had taken to Yarnell, passing through the ruins of the town they tried in vain to save. On the night of June 30, more than a hundred homes had burned, but miraculously, no civilians were killed.

The Central Yavapai Fire District named Gary Cordes the Firefighter of the Year for his actions at Yarnell. When the procession passed through town thirty-six hours later, there was still heat in the ashes in the basin where the nineteen men died, but the Yarnell Hill Fire would grow by only two thousand more acres before it was officially declared contained on July 10.

Well-wishers and first responders lined Highway 89 all the way from Yarnell to Prescott. Fire engines pushed their ladders into a triangle over the road—another tunnel like the one the men had formed for Renan that day—and the hearses drove through it as silence fell over Whiskey Row.

At the Yavapai County Medical Examiner's Office, a smoke-jumping plane flew overhead and dropped nineteen purple ribbons, one hotshot's name written on each, over the hearses. Donut followed the procession in Granite Mountain's chase truck. At the courthouse, along with four other firefighters, he helped unload the gurneys into the morgue. The bodies of his nineteen friends lined the walls of the cold and sterile room. Donut asked to be left alone with them.

Inside, it was thirty-six degrees—cold enough to slow decomposition, cold enough to see his breath. He could see a few of the men's feet, blackened and so very naked without their boots, but blankets covered the rest of their bodies. Donut was grateful for that. He didn't move from the doorway. He stood there quietly for a long time, looking out over the men until the tears came.

Donut hadn't thought of what he was going to say—what could he say? So over and over again he said what he felt. He was sorry that he wasn't there with them, that he was the one who had lived.

Nine days after Granite Mountain's deaths, on July 9, the City of Prescott held a memorial service for the men at a small stadium in Prescott Valley. Vice President Joe Biden came. So did Arizona governor Jan Brewer, Arizona senator John McCain, and Janet Napolitano, the secretary of homeland security and the state's former governor. The state fire marshal was there, as were the head of the Forest Service, fire chiefs from around the country, and more than two hundred bagpipers who played a rendition of "Amazing Grace." A total of eight thousand firefighters from 220 different agencies came.

Steve Emery wasn't there. He'd taken a trip to the mountains of Colorado to cope with the fresh and painful memories. As he'd anticipated, he couldn't get the images of the nineteen out of his head. He and the other men who had picked up their bodies agreed to meet regularly to talk about what they had seen that day. It was helping. He was functioning, managing the memories, but he was seriously considering giving up wildland firefighting. He couldn't handle another Dude Fire or Yarnell Hill.

Pamphlets describing each man in two hundred words were distributed to the crowd, along with purple ribbons that everybody in attendance pinned to their lapels in remembrance. Outside, thousands of people who couldn't find seating in the full stadium gathered to watch the memorial on giant screens. The ceremony lasted two hours.

Darrell Willis spoke, echoing a sentiment he would share many times in the coming months. "I would have followed them blindfolded into every place they went," he said.

"I didn't have the privilege of knowing any of these heroes personally, but I know them," Joe Biden said. He told a story of firefighters saving his sons' lives. In 1972, his wife, daughter, and two boys had been in a horrific car accident, and they pulled his sons out of the burning car, but his wife and one-year-old daughter didn't survive the crash. "Oh, I knew those selfless heroes," Biden said of the hotshots.

What gave Donut, Bunch, Maldonado, and Renan immense pride was that a representative from each of the 114 hotshot crews in the country came to pay their respects. It was what Marsh would have wanted. The hotshots attended the ceremony in their simple uniforms: a cotton T-shirt, green Nomex pants, and a pair of worn boots that most, as a sign of respect, had freshly oiled.

Donut spoke near the ceremony's end. On his way to the podium, he hugged Biden, Brewer, and Darrell Willis. It took a moment for Donut to compose himself at the mic.

"My brothers, my sisters, my family, I'd like to share the Hotshot Prayer with all of you."

Then he calmly read:

> *When I am called to duty, Lord,*
> *To fight the roaring blaze,*
> *Please keep me safe and strong:*
> *I may be here for days.*
> *Be with my fellow crew members*
> *As we hike up to the top.*
> *Help us cut enough line*
> *For this blaze to stop.*
> *Let my skills and hands*
> *Be firm and quick,*
> *Let me find those safety zones*
> *As we hit and lick.*
> *For if this day on the line*
> *I should answer death's call—*

Donut's voice started to crack—

> *Lord, bless my hotshot crew,*
> *My family, one and all.*

"I miss my brothers," Donut said. He walked offstage to a standing ovation, most of the audience in tears.

EPILOGUE

Over the next four months, ninety-five hundred items of tribute to the men were set by the chain-link fence outside Granite Mountain's station: T-shirts (1,100 of them), flags (900), fire helmets (20), stuffed animals (31), Pulaskis (10), letters and cards (4,000), a picture of an ice sculpture wearing turnouts, and a large wood carving of Granite Mountain's logo.

For months after the tragedy, at every hour of the day, visitors came to the station either to leave something or just to look. There was a lot to take in. So many items were hung on the fence that at one point, after a heavy rain, it collapsed under the weight.

The city didn't know what to do with the shrine until a pair of retired widows volunteered to preserve it. On September 10, during a ninety-minute lull in the rain, which had been falling steadily for four days, thirty volunteers and forty firemen took down the saturated shrine. Each section of the fence was photographed, then the items removed and loaded into a box. Two trucks, one of them provided by the Forest Service's Prescott Hotshots, were filled. Over the course of the next months, volunteers dried out and cataloged every item. One of them is a child's fire helmet with a handwritten letter folded up and tucked inside. Another, item 2013-27-57, is a close-up photo that shows a Granite Mountain widow holding the blackened hand of her husband. The hope is that someday, every item will be displayed in a museum for the men. People still leave items at the fence.

Donut received countless gifts as well. The never-ending stream of letters, shirts, trinkets, and even gifts for his daughter, though absolutely well-intentioned, eventually amounted to an outpouring of compassion that he couldn't possibly process. After the memorial, McDonough was thrust into the national spotlight. He went to New York City to appear on the *Today* show and *60 Minutes* and did dozens of interviews for local, national, and international publications. He was the guest of honor at the Dierks Bentley benefit concert for the fallen. When he was brought onstage, girls in spaghetti-strap tank tops screamed in unison, "We love you, Donut!"

He'd promised his friends they'd never be forgotten. But the reasons he spoke publicly about the event were simpler. His story could help raise money for the families of the men he loved.

"Yarnell was world news and media exposure is the single biggest driver of fundraising," says Burk Minor, the managing director of the Wildland Firefighter Foundation, a nonprofit that raises money for the families of fallen wildland firefighters. "Most deaths on the line are like Luke Sheehy, the smoke jumper who died in backwoods California in June 2013. There's no media and very little money raised. The families of Granite Mountain got exponentially more than those of any fallen wildland firefighters in history."

The Dierks Bentley concert alone raised $476,000 for the families of the nineteen, which was still only a fraction of the $13 million donated to Granite Mountain by year's end.

After funerals and worker's compensation benefits, the tragedy is expected to cost the City of Prescott $51 million over sixty years. After much heated debate, the city council voted not to reinstate a hotshot crew. Granite Mountain would be no more. Darrell Willis was devastated. What had been one of the country's most progressive fire departments was now a cautionary tale for every western city in the country. Tucson, home to the only other municipal hotshot crew in the nation, had founded that program using ideas first implemented in Prescott. Though the city denies that Granite Mountain's deaths had anything to do with the decision, Tucson disbanded its hotshot crew within a year of Yarnell Hill.

The only bright spot is that the fuels-crew-only model that Willis and Marsh helped establish is being used in a number of towns throughout the West. Santa Fe and Boulder, Colorado, both funded thinning and chipping crews to prepare their towns for the inevitable wildfires. Every year, similar initiatives are being funded by cities all across the West. Though the vast majority of western towns are still unprepared for future fires, as Yarnell was, that's beginning to change.

Two official investigations studied what happened to Granite Mountain on Yarnell Hill. The first, compiled by a fifty-member interagency team of subject-matter experts that ranged from meteorologists and historians to retired Type 1 incident commanders, found "no indication of negligence, reckless actions, or violations of policy or protocol." The second, a smaller investigation by the Occupational Safety and Health Administration, fined the State of Arizona $559,000 for four key safety violations, including implementing "suppression strategies that prioritized protection of non-defensible structures and pastureland over firefighter safety," when the state knew that "suppression of extremely active chaparral fuels was ineffective." OSHA pointed out that if Brian Frisby hadn't arrived on his Razor at just the right moment, Donut would have been Yarnell Hill's first fatality. But the question both investigations failed to answer was the one everybody wanted to know: Why did Granite Mountain leave the black?

One City of Prescott firefighter was so confused by what happened that he took an arson-sniffing dog to the fire site. The dog found the remains of a fusee. Some took it as evidence that somebody, perhaps the Helms, had lit an intentional backfire that had trapped the hotshots. But with fifty-mile-an-hour winds funneling into the basin and already extreme fire behavior raging, it's very unlikely that such a backfire, if one was even set, would have been responsible for the tragic outcome.

Independent journalist John Dougherty dedicated most of a year to examining the enduring mystery. He pressed both investigation teams to make public all of their raw research material. Eventually, officials from both teams agreed to release interview transcripts, videos, photos, and maps to the public. The archives, though redacted, are incredibly comprehensive.

A message board on Dougherty's website became the de facto forum for Yarnell Hill buffs; it swelled to hundreds of thousands of comments, mostly from anonymous citizens and firefighters obsessed with the mystery surrounding the tragedy.

Dougherty and many others turned up an astonishing number of clues the investigative team had missed, left out, or possibly just ignored in their rush to release a report. Included were a few video records that contradicted the investigations' claims that there was a thirty-minute communication gap between Marsh and the incident commanders. There wasn't. Marsh had talked to Burfiend fifteen minutes before the burnover. Burfiend, fielding countless radio transmissions during what's been described as the fire's complete audio chaos, had simply forgotten the conversation and didn't remember it until days later, when he was shaving.

This discovery, and others like it, provided more fodder for conspiracy theorists. Most people combing through the raw materials—or at least most of those posting on the message boards—seemed to be doing so in the belief that both investigation teams had willfully ignored the truth in an attempt to cover up somebody's mistakes. Whom they accused of making the mistakes depended on the commenter's perspective or agenda, and many of the most popular theories evolved with the research.

Some speculated that the incident commanders and Darrell Willis had ordered Marsh and the crew back to Yarnell for structure protection. Others blamed tension between Marsh and Steed for the fatal decision. One rumor emerged that the captain and the superintendent had a blowout fight on the saddle above the basin. The fight supposedly ended with Marsh, who was theorized to be at the Helms' place or moving back up the basin toward the hotshots, ordering Steed to bring the crew down against his will.

As of yet, no radio transmissions or other evidence have surfaced to support the theory that a disagreement occurred, let alone a full-blown argument. Marsh's location, and precisely what prompted him and Granite Mountain to leave the black, remains unknown. In all likelihood, the crew, seeing the flames bearing down on Glen Ilah, left the safety zone of their own volition, to reengage the firefight. Even after confirming receipt of the weather updates, Steed and Marsh had most likely underestimated the intensity of the wind shift, its effect on the fire, and the roughness of the terrain between the two safety zones. In the aftermath, some firefighters saw themselves as capable of making the same decision—such was the complexity of the environment at Yarnell Hill—while others saw it as an egregious and unforgivable error in judgment.

Most experts who studied the fire agreed that the decision to leave the black was ultimately made by Marsh. Jim Cook, the founder of the Wildland Fire Leadership Program, explained it this way: "As the division, Marsh was holding all of the cards. He had the power to refuse any assignments he didn't like, and Steed would have needed to be exceptionally confident to flat-out refuse Marsh's order. That would fracture a crew."

Many firefighters and community members found the conclusions of the two investigations to be, in general, deficient. Twelve families of the deceased banded together to sue the State of Arizona, the Arizona State Forestry Division, the Central Yavapai Fire District, and Yavapai County for wrongful death. Listed on the claim were Todd Abel, the operations chief; Russ Shumate, the first incident commander; and Roy Hall, the incident commander during the fatalities. City of Prescott officials were conspicuously absent from the claim.

In addition to $10 million for each surviving widow, $7.5 million per child, and $5 million for each parent, the families requested an injunction requiring the state to equip all firefighters with GPS tracking devices and personal fire shelters capable of withstanding the two-thousand-degree fire that killed the men in the basin. These shelters don't yet exist in a portable size. The lawsuit, which will likely take years to resolve, also demanded that Arizona create a memorial to the crew at the State Capitol.

Some families are less concerned with lawsuits and assigning blame; they're simply looking for closure. Linda Caldwell—Bob's mother and Grant's aunt—wanted to know everything that happened on the line that day. She asked for special permission to see her boys' burned bodies. She felt Bob's face through the American flag covering his body, and when she visited the deployment site, she got down on her hands and knees and collected bits of fire shelters left in the ash.

Some, like Clayton's wife, Kristi Whitted, didn't want to know anything at all. She took comfort in knowing that Clayton was in heaven and refused to put anything in her mind that she could never get out. Still others dealt with their grief by continuing to examine and dissect all the variables associated with the crew's deaths.

Travis Turbyfill's father, David, blamed his son's death on the shelters. Had their shelters been more sophisticated, in his calculation, Travis and the others would have survived. The shelters failed because they were exposed to conditions that far exceeded what they were designed to handle. Turbyfill pressured the Forest Service to review and improve its current system, which the agency is committed to doing over the next four years.

It's possible a redesign will emerge from this incident, but federal agencies are putting more energy toward leadership and decision-making training than into technological advancements. Along with Granite Mountain's nineteen fallen, fifteen other American wildland firefighters died on the line in 2013—more than in any year since 1994.

"When studying these incidents, the questions we asked were: Where are the new frontiers? Where can we really make a difference?" says Tom Harbour, director of the Forest Service's Fire and Aviation Management program. "As we talked, it wasn't building a better fire shelter, a better Pulaski, or a smaller, more powerful chainsaw. It was—standing there on the line, aircraft above us, engines below us, fire coming toward us, and rapidly changing dynamic weather—how do we process all this information with these wonderful brains we've been given? Perhaps the next great advance for our wild and prescribed fires isn't necessarily in technology, but in how we as human beings interact within the systems of natural and prescribed fire."

In addition to such enlightened pronouncements, Harbour and other agency directors have engaged in discussions about reevaluating the way land-management agencies mitigate, prevent, and contain wildfires. In July 2014, Senator John McCain introduced legislation requiring that at least half of what's spent each fire season on suppression also be spent on hazardous-fuels reduction projects. Currently, only $201 million of the $4.1 billion the Forest Service requested in its 2014 budget was allocated to hazardous-fuels reduction around rural communities.

"We need to rethink the practice of throwing billions of taxpayer dollars at wildfires year after year and begin aggressively treating our forests," said McCain. As of this book's publication, the bill has not passed Congress. If it, or one like it, eventually does, it will still take many generations before America's forests adapt to the current state of wildfires.

For the nineteen's closest friends and families, the year after the men's deaths was filled with pain and readjustment. In February 2014, eight months after Yarnell Hill, Donut stepped down from the fire department. He was battling PTSD and, as is common with people dealing with the disorder, had turned for a time to alcohol. He took a job as an ambassador with the Wildland Firefighter Foundation, where he spoke about his experience at benefit events for fallen firefighters. Donut's personal goal is to build an actual center in Prescott for first responders coping with PTSD. He has a blueprint, and he's on the board of an organization working to achieve the same goal, but mostly he's focusing on healing.

"I don't see it getting better until I hit rock bottom," Donut said months after Yarnell Hill. "I just can't seem to get the negativity out of my mind."

Some family members of the other hotshots publicly questioned Donut's role in the Yarnell fire. Though asked by investigators multiple times, Donut has never made a direct public statement on what specifically Marsh and Steed were discussing on the crew's private network before leaving the black. In the aftermath, "Nineteen guys made that decision" became his and the City of Prescott's mantra.

"I will make sure that it's known that there was no bad decision made," Donut said. "That no one is at fault for what happened. Anyone that does come out with negative thoughts, I'll make it known that I was there. And I know what happened. And it was just an accident. These things happen."

That wasn't enough for some families.

"He was their brother, and he owes it to them [and to those of us] trying to wrap our heads around it to start speaking up," said Juliann Ashcraft, Andrew's widow. She said her husband "would tell me as he laced up his boots every day, 'They tell me *Jump,* I say *How high?* I love this community and love serving them with all I've got.' I've been underwhelmed and upset by Donut's actions and his lack of answers."

The other surviving hotshots have tended to avoid the controversy as much as possible. Renan Packer went back to parking cars at the Scottsdale golf course. He continues to apply for structural firefighting jobs. Bunch couldn't handle the memories of Granite Mountain everywhere he looked, and he left Prescott for an arborist job in Seattle shortly after the deaths. But the dampness of the Pacific Northwest, plus black mold in his rental, threatened to make his new baby sick. The Bunches moved back to Prescott just a few months later, and in the spring of 2015 he planned to return to hotshotting.

Heather Kennedy and Scott Norris's family did not join the lawsuit. In their estimation, Scott had understood the risks he exposed himself to, and they didn't believe a lawsuit would honor his wishes. Heather had the hardest time believing that he hadn't spoken up when the crew decided to leave the black. "He knew better," she said.

Heather stayed with the Prescott Police Department and bought a house, as she and Scott had planned on doing together. When she moved, among the last things she took from their home was the chewing gum Scott had left in the shower on one of their final nights together. For months after Yarnell Hill, she sent him text messages. Sometimes they were updates about her life and the dogs. Mostly, they were pained. On August 20, almost two months after Scott died, she texted him:

"Sometimes I feel as if everyday is just another step closer to seeing you again. But it's about the journey, not the destination, right?"

Leah Fine grew to disdain media coverage that cast Grant and the eighteen others as selfless heroes, friends to the end who died proudly doing the job they'd dreamed of. She knew that if Grant had fully understood what he was risking, he never would have signed up for Granite Mountain in the spring of 2013. He'd still be washing dishes at the Mexican restaurant while trying to find a paramedic's job. And he would have been content to be at home with her. Leah, though, now had less reason to stay home, and in the fall after Grant died, she decided that adventure and hard work might help her cope. She decided to become a hotshot.

ACKNOWLEDGMENTS

First and foremost, I owe a great debt of gratitude to the people who shared with me their own painful experiences and memories of the men they loved. I sincerely hope the book does justice to the stories passed on to me by the hotshots' friends and family members who graciously gave me their time, the following in particular: David, Linda, and Claire Caldwell; Leah Fine; John, Jane, and Amanda Marsh; Wade Ward; Darrell Willis; Phillip Maldonado; Pat McCarty; Marty Cole; Brandon and Janae Bunch; Jeff Phelan; Todd Abel; Truman and Lois Ferrell; Steve Emery; Conrad Jackson; Karen and Jo Norris and Heather Kennedy; Kristi Whitted; Brendan McDonough; and Renan Packer. These people provided the book's narrative—its heart.

I also wish to thank those who told me their stories of working with Granite Mountain in the 2013 fire season and helped me understand the larger world of wildland firefighting. An incomplete list: Jim Cook, Rick Cowell, Stan Stewart, Travis Dotson, Mark Linane, Greg Overacker, Beth Melville, Tirso Rojas, Jason Schroeder, Kristen Honig, Allen Farnsworth, Dan Bailey, Dave Provincio, Carrie Dennett, Park Williams, Chuck Maxwell, John Wachter, Fred Schoeffler, Josh Barnum, Todd Haines, Todd Lerke, Harry Croft, and James

Lewis. I owe a personal thanks to Jennifer Jones, the Forest Service's public information officer, who was incredibly accommodating during both the book's initial reporting and its fact checking. These people provided valuable information that conveyed the greater context in which the tragedy occurred.

Personally, I must also thank a long list of people who helped produce this book. For starters, thank you to my wife and best friend, Turin. Your endless patience and support made this possible. I thank my parents, Bonnie and Paul, who taught me to love books, and my older brother, Garrett, who has been the first editor of nearly every piece I've written since we were kids. Sam Moulton read the manuscript cover to cover multiple times, and his sharp edits and steady encouragement exponentially improved the book. Thank you to my friends Peter Vigneron and Frederick Reimers, who read countless early drafts. Dave Costello, Abe Streep, Sean Cooper, Jonah Ogles, and Grayson Schaffer all provided thoughtful input on various chapters. Jakob Schiller provided many of the images in the book and also helped me report the story by sharing with me many of the initial contacts he'd made while photographing the crew in 2012. Thank you. Chris Keyes edited the National Magazine Award–nominated feature that was the seed for this book; Lorenzo Burke published it in *Outside;* Alex Heard served as a general counsel throughout the entire process; Kevin Fedarko's input from outline to final draft was invaluable; and my remarkable copy editor, Will Palmer, added more commas than any man should ever have to and did so while providing poignant and necessary feedback on the story's bigger picture. Reid Singer factchecked every page of the book—a monumental effort he carried out with poise. He contributed greatly to the book. And finally, thanks to my excellent agent, Jennifer Joel, and editor, Mark Tavani, who offered clear edits and shepherded me through every step of the long process.

NOTES

The pages that follow document the main sources used in each chapter. I have not listed the source of every quotation, fact, or passage, but to ensure the reporting's accuracy, the entire book was independently fact-checked. Descriptions of fire behavior and the sights and sounds of the blazes Granite Mountain fought in 2013 were informed by interviews, photos, and videos shot by fire-fighters on scene and by my own time on the fire line. Most information specific to the personality of Granite Mountain's 2013 crew came from the three surviving members: Brendan "Donut" McDonough, Renan Packer, and Brandon Bunch. In total, I spent nearly a month reporting the book in Prescott. This time included countless cups of coffee at the Raven, a café where I met many sources, and multiple trips to the fire site at Yarnell. I also visited the Doce Fire site with McDonough and the Thompson Ridge Fire site with Todd Lerke.

PROLOGUE

Details about the Yarnell Hill Fire of June 30, 2013, came from interviews with Brendan "Donut" McDonough. The timeline of events was provided by the two official investigations into Yarnell Hill—the Arizona Department of Occupational Safety and Health and the Yarnell Hill Serious Accident Investigation—as well as interviews and official transcripts of interviews with McDonough, Todd Abel, and Brian Frisby that were made public through the Freedom of Information Act (FOIA). Weather updates came from both the investigations and subsequent conversations with fire weather meteorologist Chuck Maxwell at the Forest Service's Southwest Coordination Center.

CHAPTER 1

Information about the first week of Granite Mountain's season was provided by Bunch, Packer, and McDonough. Linda Caldwell, David Caldwell, and Leah Fine provided Grant McKee's biographical information, and a number of trips to Granite Mountain's station allowed me to collect details about the saw shop, the ready room, and the setting. Former hotshot superintendents Rick Cowell, Mark Linane, Jim Cook, and Stan Stewart provided general background information on the hotshots and wildland firefighters, and this was supplemented by material provided by Jennifer Jones, the public information officer for the Forest Service. The total size of America's wildland firefighting force, a surprisingly tricky number to pin down, came from Dan Bailey at the International Association of Wildland Fire. The number fifty-six thousand accounts for city, state, and county wildland firefighters, but it's considered a conservative estimate. When volunteers are added, the number of American wildland firefighters is thought to be as high as seventy-five thousand.

CHAPTER 2

Details about the hotshots' training day were provided by Bunch, Packer, and McDonough. Prescott's Wildland Division chief, Darrell Willis, who oversaw the drill, also provided information on the hotshots' certification process. Information about Prescott's history came from the book *Prescott Fire Department* (Arcadia Publishing), by Eric Conrad Jackson, as well as visits to the courthouse square, Whiskey Row, and Prescott's Sharlot Hall Museum. Heather Kennedy and Karen Norris told me Scott Norris's story. Michael Thoele's book *Fire Line: The Summer Battles of the West* (Fulcrum), along with Jennifer Jones, Chuck Womack, and Kari Boyd-Peak, at the National Interagency Fire Center (NIFC), provided background on how the response to wildfires is coordinated. Longtime hotshot Tirso Rojas augmented my own experience on the fire line with explanations of how sawyers and swampers work. Stephen Pyne's extraordinary book *Fire: A Brief History* (University of Washington Press) proved an invaluable source for the history of wild flames in North America; Timothy Egan's *The Big Burn* (Houghton Mifflin) was the reference for Ed Pulaski's backstory; and the Interagency Fire Shelter Task Group's *Wildland Fire Shelter: History and Development of the New Generation Fire Shelter* was a touchstone for explanations about fire-shelter usage in the United States.

CHAPTER 3

Eric Marsh's backstory came from interviews with his parents, John and Jane Marsh, and his wife, Amanda. His second wife, Kori Kirkpatrick, confirmed much of his early history in Prescott. Marty Cole, Phillip Maldonado, Wade Ward, and Pat McCarty—all former Crew 7 and Granite Mountain firefighters—provided background about Marsh's mission to turn Granite Mountain into a hotshot crew, and Marsh's personnel files were a source for performance evaluations throughout his tenure on Granite Mountain. Jim Cook contextualized hotshot culture; Packer and Bunch shared anecdotes from the night the hotshots slept out after their drill.

CHAPTER 4

My sources for the state of wildfires in the West came from NIFC's daily situation report. Details about the prairie fire were supplied by multiple newspaper articles published in the Prescott *Daily Courier.* Bunch, McDonough, and Packer all shared with me their experiences on the fire. Weather information came from the NIFC-issued summary of the 2013 fire season and meteorologist Chuck Maxwell, who tracked the season's development over many months. Packer provided his own story. Bunch, Heather Kennedy, Pat McCarty, and Wade Ward all corroborated Marsh's interview practices.

CHAPTER 5

Details of Brandon and Janae Bunch's home life came from the Bunches. Former hotshots Pat McCarty and Jeff Phelan provided biographical information on Garret Zuppiger, as did Zuppiger's well-written and highly entertaining blog *I'd Rather Be Flying!* (garretjoseph.wordpress.com). Bunch, Packer, Phillip Maldonado, Heather Kennedy, McDonough, and Leah Fine all provided information contrasting the leadership styles of Marsh and Jesse Steed.

CHAPTER 6

Jennifer Jones explained the fundamental operations of NIFC. This reporting was supplemented by Brian Mockenhaupt's wonderfully clear *Atlantic* story about the Yarnell Hill Fire, "Fire on the Mountain." Region 3's public information officer, Mary Zabrinski, provided basic information on the state of wildland fires in the Southwest in early May 2013, but the primary source for this chapter was meteorologist Chuck Maxwell, whom I visited at his office in Albuquerque. His early-season forecasts mapped out the movement of the fire season. Bioclimatologist Park Williams, of the Tree Ring Lab at Columbia University, provided scientific context for the warming and drying of the Southwest. Todd Lerke told me his own story of the initial attack on Thompson Ridge, and former hotshot superintendent and longtime fire researcher Fred Schoeffler gave me the growth rate of the Las Conchas Fire.

CHAPTER 7

Heather Kennedy told me about Scott Norris and their relationship, as well as the details about the fire-and-weather video. This reporting was supplemented by conversations with Scott's mother, Karen, and his sister, Jo.

CHAPTER 8

Mike Johns, a U.S. assistant district attorney, offered invaluable context through both personal interviews and his meticulously cultivated report on the incident, "The Dude Fire." Supplemental reporting came from Jaime Joyce, in her shocking story "Burn," published by the nonfiction site the Big Roundtable on June 23, 2013. National Park Service–contracted helicopter pilot Chris Templeton gave sensory details about flying a JetRanger beside a collapsing column, and the video "Dude Fire Staff Ride" (youtube.com /watch?v=EaV5WKgKVHo), which highlighted the interaction between a number of surviving sources, provided quotes from a number of hotshot superintendents on scene on June 26, 1990. Ultimately, though, the primary source for this reporting was the official investigation into the Dude Fire fatalities and the many pages of handwritten interview transcripts associated with the deaths, which are available at the Wildland Fire Staff Ride Library (www.fireleadership.gov).

CHAPTER 9

Payson Hotshots superintendent Mike Schinstock explained his crew's educational walk-throughs of the Dude Fire site. Heather Kennedy gave background on Scott's fascination with prior burnovers and his relationship with Kevin Woyjeck. Context about how the Forest Service and other wildland fire agencies respond to tragedy fires came from Jim Cook, Rick Cowell, and sources who wished not to be named but who are employed by large, nationally funded educational institutes that study wildland fire. Norman

Maclean's brilliant book *Young Men and Fire* (University of Chicago Press) provided context for this chapter and many others in the book, as did his son John Maclean's exceptional investigation of the tragic South Canyon Fire, *Fire on the Mountain* (Simon & Schuster). Bunch, Packer, and McDonough provided details about the Hart Fire, and Don Muise of the Coconino National Forest gave the ranger district's reaction to the fire.

CHAPTER 10

In addition to trips inside Alpha's and Bravo's buggies, Bunch, the sawyer on Bravo, and McDonough, lead Pulaski on Alpha, gave me the details on the buggies' interiors. Bunch told me the story of the chips, which was confirmed by Packer and Leah Fine. FOIA requests for Incident Action Plans (IAPs) and ICS 209s—forms that document the incident commander's response to a fire—detailed Bea Day and her management team's response to the Thompson Ridge Fire. These documents also specified personnel assignments and division objectives on the fire. Bunch, McDonough, and Packer all related to me their experiences on Thompson Ridge, as did the Division Zulu, Allen Farnsworth, and the photographer Kristen Honig. Videos and photos that Honig shot provided additional scenes of Granite Mountain's experience on Thompson Ridge. Details on relative danger rates of professions came from the Bureau of Labor Statistics. Between 2000 and 2013, an average of forty-six mail carriers were killed at work per year, compared with thirty-three wildland firefighters.

CHAPTER 11

Fire researcher Jim Cook's paper "Trends in Wildland Entrapment Fatalities . . . Revisited" was instrumental in reporting about history's largest fire fatalities. So were Tim Egan's *The Big Burn,* Stephen Pyne's *America's Fires: A Historical Context for Policy and Practice* (Forest History Society), and conversations with James Lewis, at the Forest History Society. Bioclimatologist Park Williams provided context on the impact that fire suppression and climate change are having on western forests. Harry Croft gave background on the evolution of fire management policy, Alexander Evans contextualized the role of prescribed fire in the West, and Jerry Williams's writing underscored the onset of the era of mega-fires in his 2011 paper "Mega-Fires and the Urgency to Re-Evaluate Wildfire Protection Strategies through a Land Management Prism." Details about the Peshtigo Fire came from Lee Sandlin's book *Storm Kings* (Pantheon) and Peter M. Leschak's book *Ghosts of the Fireground* (HarperCollins). Steed's backstory came from a combination of reports by Josh Eells at *Men's Journal,* Brian Mockenhaupt at *The Atlantic,* Steed's personnel files, and stories told to me by hotshots he'd worked with throughout his career. McDonough told me his own story.

CHAPTER 12

Details about Marsh's first years as a hotshot superintendent were told to me by his wife, Amanda, and his parents, John and Jane. Former hotshot superintendents Jim Cook, Stan Stewart, and Mark Linane contextualized what new superintendents are often exposed to during their first few years on the job. Marty Cole, Crew 7's superintendent before Marsh, added details about Marsh's and Granite Mountain's particular experience. Marsh's personnel files were the primary source for details about the tension between Marsh and Aaron Lawson, the crew's captain before Steed, but Maldonado, Willis, and Bunch corroborated and expanded upon the tensions.

CHAPTER 13

Incident Action Plans and ICS 209 forms were the sources regarding the fire's expansion; so too was information provided by Darrell Willis, who was on the scene. Heather Kennedy showed me a video of the skit mimicking the After Action Review; Janae and Brandon Bunch told me about Brandon's last few days on Thompson Ridge; Leah Fine explained Woyjeck's relationship with Grant McKee; and McDonough told me about his trip to Las Vegas and his decision to come to God. Maldonado, who was McDonough's squad boss at the time, confirmed the story about McDonough and the crew's experience in Las Vegas.

CHAPTER 14

Details about the crew's off-days came from McDonough, Bunch, Packer, Leah Fine, Jo Norris, and Heather Kennedy. Chuck Maxwell's forecasts provided information on the weather conditions in early June. Prescott's fire history was sourced from Eric Conrad Jackson's *Prescott Fire Department* and a number of local museums in the area. Details about the Indian Fire, which nearly destroyed Prescott the year that Crew 7 was created, came from the Prescott *Daily Courier.* Willis told me some of the history of the Prescott Area Wildland Urban Interface Commission, but the organization's story was documented most thoroughly by Everett Warnock, in the paper "A History of the Prescott Area Wildland/Urban Interface Commission." Interviews with firefighters and dispatchers at NIFC, Dan Bailey at the International Association of Wildland Fire, and Harry Croft placed Prescott's challenges within the greater context of wildfires in the West. Budget numbers past and present were sourced from James Lewis at the Forest History Society and Jennifer Jones at NIFC.

CHAPTER 15

During a visit to the Doce Fire site with McDonough in November 2013, Donut told me the story about his close call and took me to the juniper, where he explained how the crew had saved the tree. Heather Kennedy, who heard of the incident from Scott Norris, confirmed the severity of McDonough's close call. The number of houses the Doce threatened, the fire size, and growth rates all came from ICS 209s and IAPs. Conversations with Incident Commander Tony Sciacca and articles from *The Daily Courier* and *Wildfire Today* supplemented this reporting. An online archive of Chris MacKenzie's photos helped fill out the moments that took place at the alligator juniper. Stories of the hotshots sleeping at the station came from Leah Fine, Kristi Whitted, Claire Caldwell, and McDonough. Packer told me about returning to the station. Chuck Maxwell and archival weather information were the sources for fire severity in the last weeks of June.

CHAPTER 16

During my visits to Yarnell, Lois and Truman Ferrell told me about watching the Yarnell Hill Fire ignite and the hours and days that followed. Russ Shumate's story came from comprehensive interview transcripts and handwritten notes from both the Arizona Department of Occupational Safety and Health Investigative Report and the Serious Accident Investigation Report. These notes, obtained through FOIA requests from the excellent independent journalist John Dougherty, among many others, also provided the foundation for all the Yarnell Hill chapters that follow. Amanda Marsh told me about her dinner with Eric Marsh on June 29, and McDonough and Jeff Bunch, the bartender, told me about the hotshots stopping to get a beer at Moctezuma's that same night.

CHAPTERS 17–23

The book *From Tragedy to Recovery: The Yarnell Hill Wildfire of 2013* (Createspace), released by the Yarnell Chamber of Commerce, provided the history of Yarnell. Details of Granite Mountain and other firefighters' experience on the Yarnell Hill Fire on June 30 came from conversations with Todd Abel, Conrad Jackson, Steve Emery, the Ferrells, and McDonough, as well as interview transcripts and notes from, among others, Abel, Russ Shumate, Byron Kimball, Gary Cordes, Paul Musser, Darrell Willis, Rance Marquez, Rory Collins, Brian Frisby, Rogers Trueheart Brown, three other Blue Ridge Hotshots, and the hikers Joy Collura and Tex Gilligan.

Both official investigations into the tragedy provided a timeline of events for the fire and details about the number of tanker drops, weather updates, firefighter movement, and the fire's rate of spread. Chuck Maxwell told me about the scene in Albuquerque and explained the weather phenomena in play that day. John Wachter also helped explain the day's weather. Heather Kennedy described the weather in Prescott and provided her text message conversations with Scott Norris. Photo and video evidence from McDonough's phone, in addition to videos recovered from Chris MacKenzie's burned phone and camera, helped me re-create the near midair collision. It's important to note that aviation experts speculate that the aircraft were in fact a safe distance apart. I relied on McDonough's telling of the incident, as well as the videos and Scott Norris's text message, because they demonstrate how the hotshots' nerves were frayed hours before the tragedy.

The number of firefighters on scene came from a comprehensive list of resources used to fight the blaze. Carrie Dennett, the state's excellent public information officer, provided this list.

Experts on Yarnell Hill will notice a slight time discrepancy between the book's version of events and the official investigation. The investigation has McDonough getting pushed off his lookout at 3:55 P.M., but time signatures on photos that McDonough shot at Granite Mountain's rigs, after Frisby picked him up, show that this event occurred twenty-four minutes earlier. As firefighters reported reliable cell service in the area, there's little reason to doubt the accuracy of McDonough's phone. Logic, too, would dictate that Steed received the fire-wide weather update at the time it was delivered (3:26 P.M.) and not twenty-four minutes after (3:50 P.M.). Cell-phone video recordings documented a number of radio conversations between Marsh and Steed that took place while the hotshots were on the ridge, and McDonough provided the details of Marsh and Steed's radio conversation about whether to leave the safety of the black.

The radio transmissions delivered moments before the hotshots died were recorded on a helmet-camera video shot by firefighters working near Glen Ilah. The flames' rate of spread and the fire behavior in the basin at the time the men were killed were determined through fire-behavior analysts whose research was included in the official investigation. Calculations based on these reports were used to estimate the distance between the flames and the hotshots during the final radio conversations Steed, Marsh, Caldwell, and Abel had with Bravo 33. Details of what it's like to deploy came from accounts by burnover survivors and from training procedures all hotshots are required to go through. A former hotshot superintendent who visited the deployment site days after the hotshots' deaths passed on the detail of a deer racing out ahead of the flames. He found the recently burned body of a deer near the site. Post-accident reports compiled by Eric Tarr and the helicopter pilot were used to re-create the scene at the fatality site, and a body-location map plus details from the Yarnell sheriff's department were used to describe the condition and location of each fallen hotshot.

EPILOGUE

Steve Emery provided me with his story. McDonough described the moments after the fatalities were reported, the funerals, and the weeks after. The scene at Mile High Middle School was described by Heather Kennedy, Karen Norris, Claire Caldwell, Bunch, Packer, Maldonado, and others. I attended the hotshots' memorial at the stadium in Prescott Valley.

Information about the tribute fence came from Dottie Morris, one of the founders of the Tribute Fence Preservation Project.

ABOUT THE AUTHOR

KYLE DICKMAN is a former editor at *Outside* magazine
and a former member of the firefighting crew known
as the Tahoe Hotshots. He spent five seasons fighting
wildfires in California. Dickman's reporting has been
nominated for a National Magazine Award. He lives in
Santa Fe, New Mexico, with his wife, Turin.

@KyleDickman

This book was set in Hoefler Text, a typeface designed in 1991 by Jonathan Hoefler (b. 1970). One of the earlier typefaces created at the beginning of the digital age specifically for use on computers, it was among the first to offer features previously found only in the finest typography, such as dedicated old-style figures and small caps. Thus it offers modern style based on the classic tradition.